Yimagine

Pandora's Potatoes

2nd Edition

October 11, 2018

TABLE OF CONTENTS

FOREWORD

MILLIONS OF YEARS of evolution produced **The Potato**—not one variety but more than two-hundred interbreeding species. Each unique plant contains a random combination of 50,000 genes that play important roles in survival but are not sufficient to *ensure* survival. Survival is a matter of chance: individuals may die when they lack the genes to overcome the stresses confronting them but the population survives, from year-to-year and forevermore, because there are always enough plants with the right combinations of genes. Interbreeding potatoes are like people. **They thrive on diversity**.

However, our society ignores this need for diversity and, instead, is very much focused on the opposite—uniformity. There are expectations, rules, and regulations to ensure *everything we do* yields predictable results. That includes the breeding, farming, and processing of potatoes. A good potato weighs about 8-oz, has white flesh and a russet skin, and produces crisp and golden-colored fries when cut and fried.

Most of our potatoes represent the 143rd generation of clones of *a single plant that grew and died in 1873*, named Russet Burbank. Each year, ten billion identical clones of this plant are grown around the country. What we call potato farming is in fact the largest and oldest experiment in mass-cloning.

About two decades ago, processors came to the conclusion that even the cloned Russet Burbank was too variable, that too much variability was encoded within its fixed genetics. Some spuds darkened more than others did; they either developed black spot bruises or turned too golden-brown when fried. This unpredictable darkening was no longer acceptable. Consumers deserved potatoes and fries that were consistently bland, pale, and blemish-free!

The easiest way to suppress Russet Burbank's fickleness was to modify its genome. "Just take out whatever causes the problems," I was told. "Liberate Russet Burbank from its rot."

I was a well-known genetic engineer of potatoes, and I was expected to create a perfectly uniform superspud. But the fact of the matter was that I didn't understand DNA. I was less capable of editing the molecule of life than the average American is capable of editing the Sanskrit verses of the *Bhagavad Gita*. I just knew enough to be dangerous.

Working blindly, through trial and errors, I eliminated potato's ability to darken; it took me more than a decade of my life to develop the pale, bland, and blemish-free GMO version of Russet Burbank that processors had asked for and that consumers and farmers seemed to want. However, I never asked myself why Russet Burbank had developed the ability to darken, or if my intrusions could negatively affect the plants.

There were other ways in which I changed potatoes. For instance, my employer was concerned about the 'unhealthy' image of French fries and potato chips, and so I tried to identify and target genes that negatively affect food quality. I initially attempted to lower calorie content and fat absorption by modifying starches, or to reduce salt requirements by increasing glutamate content, but these changes proved too complex and difficult. Soon, I gave up attempting to address real, complex issues and convinced myself to focus on a self-made problem that I was confident I could solve. Normal French fries are carcinogenic, I told myself, and I will prove the value of genetic engineering by developing less-carcinogenic GMOs.

I had never visited a farm and I didn't care to know about potatoes or farming but I knew that yields were unpredictable. It was another example of potato's capriciousness: the crop was very sensitive to diseases, pests, and abiotic stresses. But it seemed I could ensure yield uniformity by incorporating resistance against late blight, potato's most important disease. I should have known better, though, because late blight is only one of the many biotic issues, and each of these issues is too complex to be overcome by simple gene manipulations.

By always ignoring my gut feelings, I believed that I had created a healthier and more predictable crop for the struggling potato industry. However, as my modifications were welcomed by the scientific community and approved for commercial use by the USDA and FDA, I sensed a growing reluctance in myself to move forward. I vaguely understood there might be hidden issues.

Forcing myself to overcome my biases and assumptions, I became aware of the first errors. They were minor but freaked me out. Could I have made more mistakes? My suppressed doubts turned into confusion and I realized that, perhaps, I had been blind. I had been trying to edit DNA, the language of life, without understanding the intricate relationship between DNA and life. All I had done was create a mirror image of my narrow-minded and distorted intellect.

I tried to slow down my program and re-evaluate what had been created, and I also began reallocating resources to other, simpler methods that were not as error-prone, such as genetic imprinting and diversity breeding. But it was too little too late. A hedge-fund manager and a statistician were involved now and, dreaming of billion dollar profits, they wanted to speed-up the effort, not slow it down. The GMO potato was out of my hands and rushed to the market.

Early in 2013, I left the laboratory with a very strange feeling in my guts, and I relocated to a small farm in the mountains of the Pacific Northwest. It was calm there and beautiful, and my corporate past seemed like a bizarre and convoluted dream. What had I been doing so blindly during all those years? Questions came up during the day, and I often spent the nights reviewing my old publications and trying to find answers. I became aware of some of the hidden issues of my work—the so-called unintended effects. These issues had to be clarified and communicated before I could truly take distance from my work.

Without understanding all the details yet, I tried to discuss my concerns with my ex-employer. But there was no interest in inconvenient truths, just like I had ignored critical feedback in the past, and I was simply reminded of the terms of confidentiality agreements. These agreements prevent me from talking about what happened behind the doors of GMO laboratories [which is fine with me—what happened there should stay there] but they don't apply to the new insights I gained by re-evaluating data in the public domain.

I thank Diana Reeves, Jonathan Latham, Jennifer Berman Diaz, and Heather Holloway for feedback and support.

Disclosures

This booklet is a work-in-progress and the text is intentionally brief. For questions or comments that might help me further clarify my concerns, please email yimagine@gmx.com. All statements are based on data and insights in the public domain, as well as on communications not covered by confidentiality agreements. They are made to the best of my abilities, and not always confirmed experimentally. I hope my assessment will contribute to a better understanding of the potential risks and issues associated with GMO potatoes.

I wrote this book in the 'I-form' but that doesn't mean I worked alone as genetic engineer. I was assisted by molecular biologists, plant cell biologists, biochemists, plant pathologists, and agronomists.

INTRODUCTION

The Main Issues of the Potato Industry

Issue 1: Vulnerability and Loss.

Russet Burbank is the oldest, most-loved, and most-hated agricultural variety in the United States. These potatoes have been cloned from year-to-year and for over 145 years, and they are grown extensively all around the country, primarily for baking and frying. But they are not perfect and growing them is not for the faint of heart: the variety is much more vulnerable to disorders, pests, diseases, and stresses than any other major crop. Farmers have to make educated guesses to protect their crop against the most aggressive pests and diseases, but this protection is expensive and never all-inclusive. About fifteen percent of potatoes still deteriorates before harvest, whereas a similar percentage gets affected after harvest (mainly because of bruise-associated disease and dehydration). Furthermore, at least five percent of potatoes that make it to the processor gets discarded in the plant and another five percent is considered useless because of too much fry-induced darkening.

Issue 2: Public Concerns About Fries.

The mostly blemish-free potatoes that end up as French fries contain large amounts of salt and fat. This *comfort food* was in high demand until the mid-1990s, when people consumed over 40-lbs of fries per year, but it has since been down-graded to *junk food* and the average consumption is below 30-lbs. and declining still.

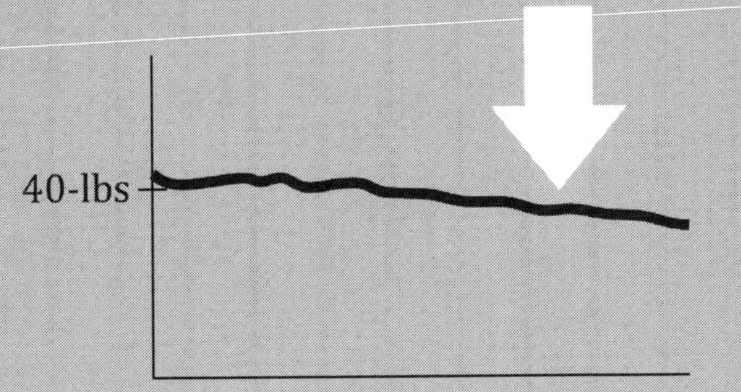

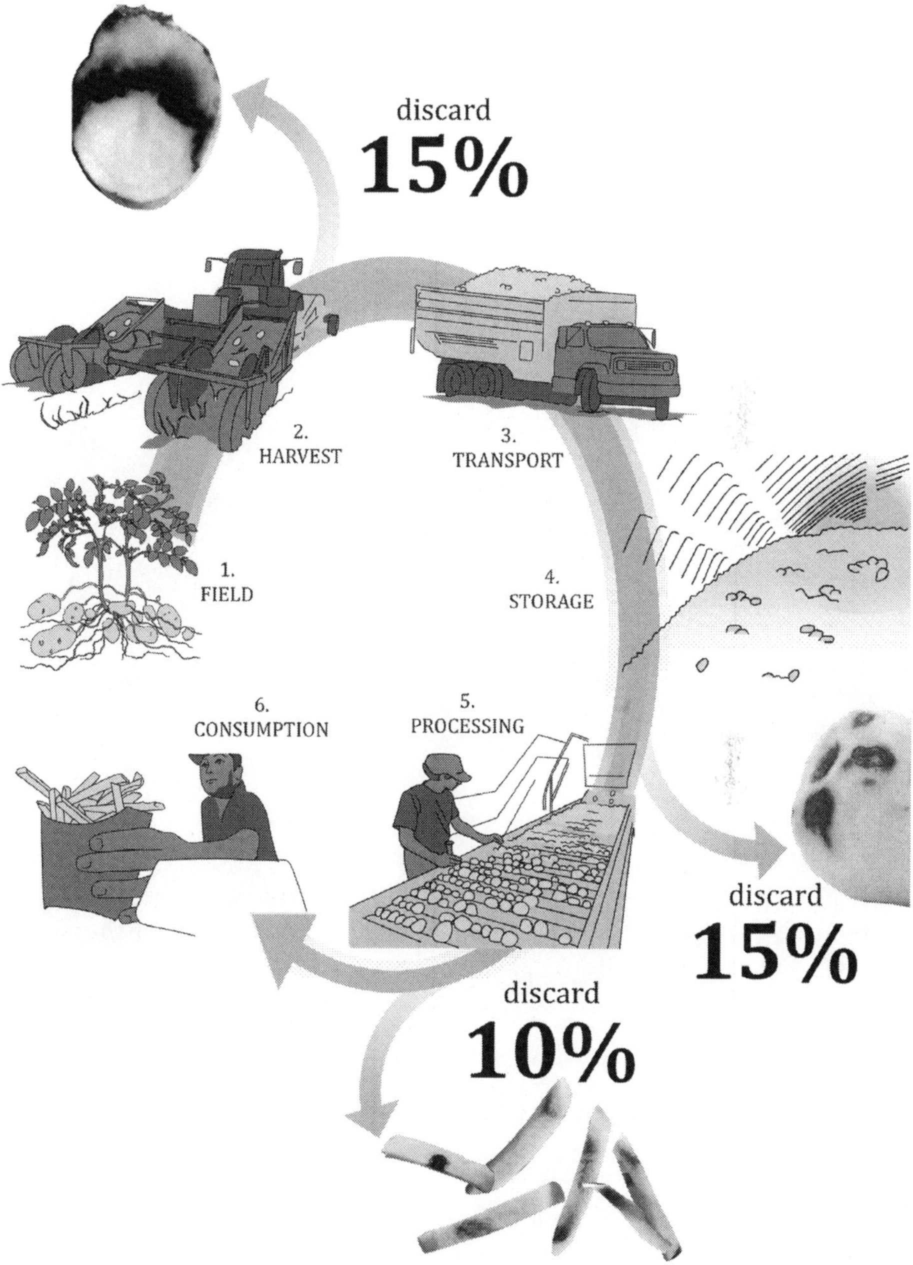
discard
15%
2.
HARVEST
3.
TRANSPORT
1.
FIELD
4.
STORAGE
6.
CONSUMPTION
5.
PROCESSING
discard
15%
discard
10%

Potato processors are squeezed between farmers and vendors. Their power is marginalized as they face not only increasing levels of competition but also declining sales volumes. Survival requires an unwavering commitment to two goals: increase efficiency and lower costs—it is the curse of any commodity business...

...**unless** processors find a way out and gain control over farmers, competitors, and vendors. Such a change of the status quo can be achieved by *solving* the main issues of the industry and *owning* the solution. In other words, processors will become more powerful if they develop proprietary potatoes that (1) are not as vulnerable as normal potatoes, (2) are more uniform and efficient, and (3) process into fries that are perceived healthier than normal fries. The easiest way to develop such a variety is by hiring a genetic engineer to (finally) change the genetic make-up of Russet Burbank.

This is where I came in.

I was supposed to modify Russet Burbank so that access to it would become a *prerequisite to success* for farmers, processors, and vendors.

My modifications were four-fold. Two of the changes were meant to prevent the development of disorders that causes the blackening of potatoes and the browning of French fries:

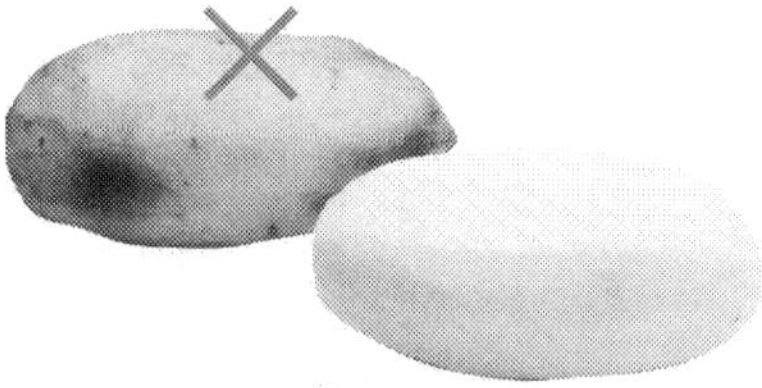

Potatoes develop black spots when they bruise, which occurs very frequently, so I silenced the blackspot bruise gene **PPO**.

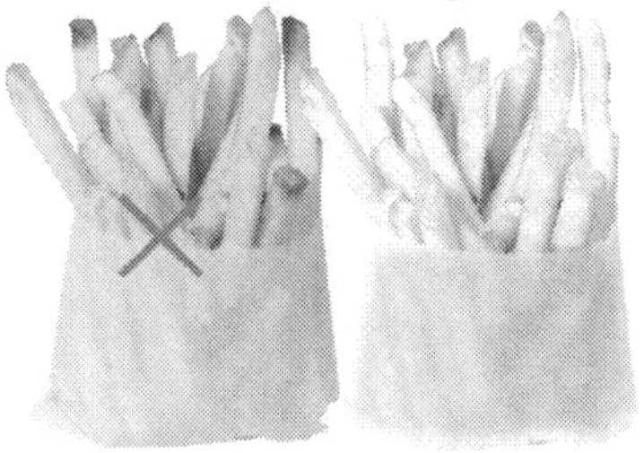

And potatoes may also brown (to some extent) when they are stored and then fried. This browning reaction is blocked by silencing the browning gene **INV**.

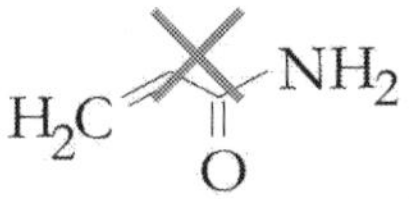

Furthermore, I believed to have improved the healthfulness of Russet Burbank by silencing the gene **ASN**, which is normally involved in the production of the carcinogen acrylamide.

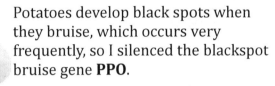

The fourth change was accomplished by introducing the **VNT** gene of wild potatoes into Russet Burbank. This **gene provides resistance against potato's most important disease—late blight.**

Early versions of the modified Russet Burbank were released under the names Innate Potato, White Russet, and Hibernate, and more sophisticated forms will be released soon.

After I left the laboratory and reconsidered my past work, I begun to realize my mistakes, and I understood that my potatoes are not what they appear to be.

They may spread sickness, death, and other evils when released, just like ***Pandora's opened box****, and so I name them Pandora's Potatoes.*

Pandora's Potatoes are claimed to be resistant to darkening and late blight and to be healthier for consumers, but the fact of the matter is that they are unintentionally designed to **conceal** infections and bruises, and that they may cause an increase rather than decrease in overall disease pressures. Furthermore, **hidden** mutations and numerous **ignored** effects of gene silencing have a negative effect on yield, whereas the bogus claim of reduced carcinogenicity **distracts** from the fact that the potatoes may contain **hidden** toxins that do not promote the health of consumers but **undermine** it.

Silencing = *Reverse Evolution*

Gene silencing is used to block the activity of specific genes that, supposedly, have a negative effect on quality and play no positive role in agronomic performance. The method is promoted as an accurate means to correct 'evolutionary mistakes'. But is it possible for evolution to make mistakes and maintain *meaningless* genes? And does engineered silencing affect the targeted genes *only*? The answer to both these questions is—**No**.

Evolution has fine-tuned the genetic make-up of the potato, not just regarding stress tolerance and yield but also in terms of the crop's nutritional value. Potato's genes should, therefore, be expected to play important roles in agronomy and quality, especially when conserved—as is the case with PPO, INV, and ASN* (these genes are so important that they are also conserved in most other plant species, as well as in bacteria, fungi, and animals).

PPO-silencing causes the deterioration of potato's natural stress tolerance and food quality (see **Chapter 1**).

ASN-silencing in roots and tubers limits the plant's capabilities to assimilate nitrogen and to use nitrogen fertilizers efficiently [1,2] (**Chapter 2**).

INV-silencing blocks the formation of the signaling molecules glucose and fructose, and so affects field emergence, fertility [3] and other agronomically-important traits, while also compromising potato's sensory profile (**Chapter 3**).

Silencing was engineered by inserting three artificial *knots* into the genome of Russet Burbank. Each of the *knots* consists of a natural gene tied together to its mirror image. Nature tends to reject these *knots*, and so it took some persistence to force them into potatoes.

---

* *As well as two additional genes, PHL and R1, that had been silenced in Pandora's Potatoes but have sinced reversed to normal, probably because of promoter methylation (Chapter 5).*

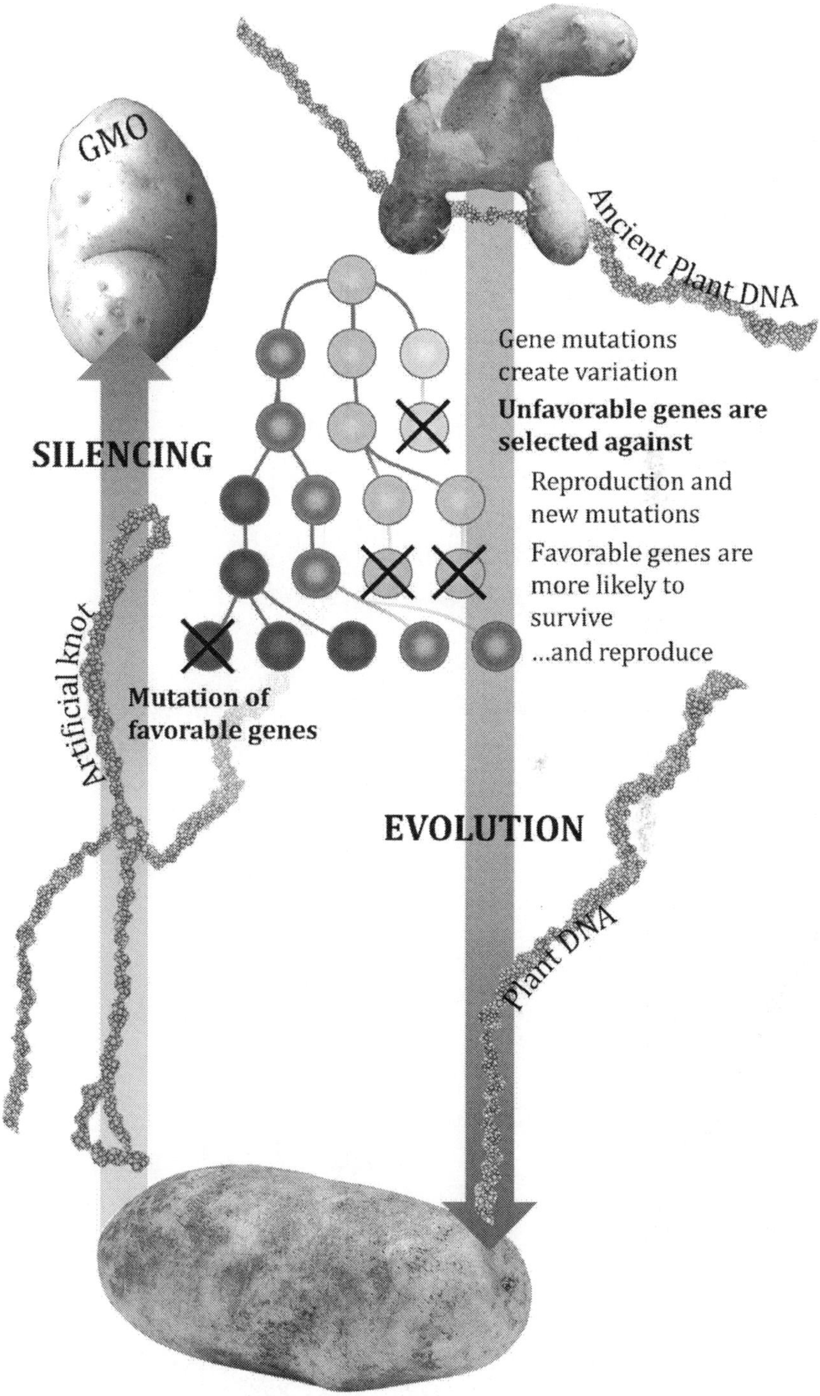
GMO
Ancient Plant DNA
Gene mutations create variation
Unfavorable genes are selected against
Reproduction and new mutations
Favorable genes are more likely to survive
...and reproduce
SILENCING
Artificial knot
Mutation of favorable genes
EVOLUTION
Plant DNA

Introduction of the *knots* triggered a panic reaction in potatoes that caused the shut-down of genes with sequence homology (PPO, ASN, and INV), like how the tied-together laces of two shoes make it impossible for a person wearing the shoes to walk**.

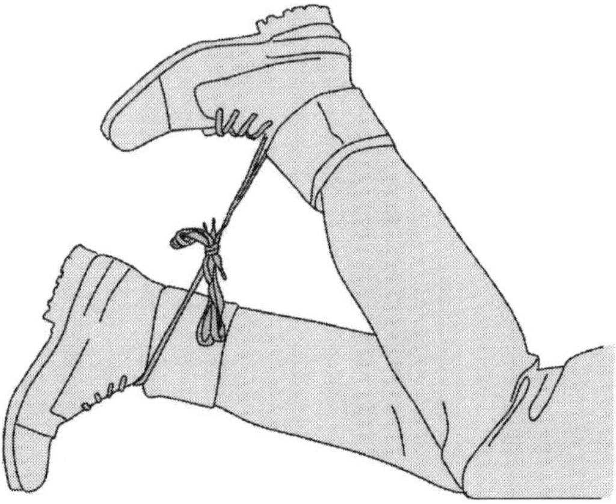

But the silencing method used to create the *knots* is different from standard silencing methods in that it was not carefully evaluated for unintended side-effects [4]. Instead of creating uniform double-stranded RNAs of expected size, sequence, and structure, the *knots* produce uncontrolled, unpredictable, and disordered assortments of double and/or single-stranded molecules that range in size from dozens to thousands of nucleotides and vary from cell to cell. Alarmingly, the silencing constructs have been described as: **"cluster bombs that, when detonated, produce a hazy cloud of diverse RNAs."**

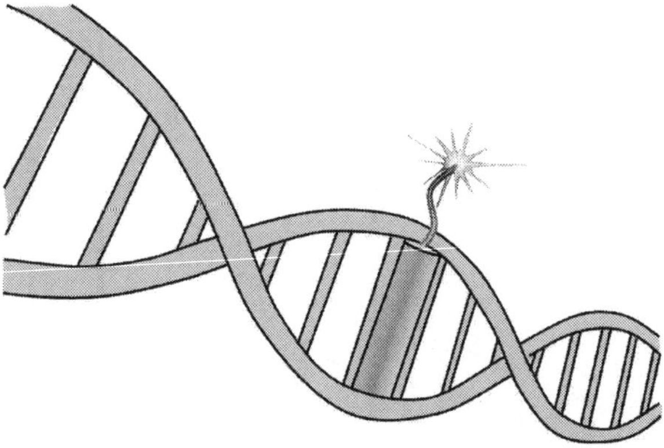

** *Genes can also be inactivated by CRISPR technology, which is more gene-specific but less tissue-specific and, consequently, about as damaging to plants as the method employed to create Pandora's Potatoes.*

The introduced cluster bomb *knots* are more effective than conventional silencing constructs [5] and they silence not just PPO, ASN, and INV but any gene with short stretches of sequence identity***.

*** *Examples of potato genes that may be silenced in Pandora's Potatoes are defensin (CV506923), 60S ribosomal protein (JZ167968), CAX-interacting protein (CK861375), aspartate-tRNA ligase (JG557979), and mitochondrial outer-membrane 'porin' (CN462452).*

1

PPO SILENCING

Concealing Bruise

SILENCING OF THE PPO GENE was supposed to have one effect only—it would stop bruise without changing anything else.

But this assumption is naive and wrong. A simple tetrazolium test shows that **Pandora's Potatoes bruise at least as much as normal potatoes**. Indeed, it is impossible to prevent potatoes from bruising, which happens all the time in the field as well as during harvest, transport, storage, and processing. Instead of developing a bruise-free GMO potato, a trick was applied to *conceal* the accumulating bruises.

Most farmers and processors don't realize that PPO-silencing conceals rather than prevents the bruising. It doesn't even matter to them. All that matters is that it is not so important anymore to be very careful during the harvest, transport, and storage of potatoes. And there is also less need for quality control: Pandora's Potatoes are not inspected for invisible bruises, and invisible bruises are not trimmed from the potatoes and fries. This *laissez-faire* approach could save farmers and processors hundreds of millions of dollars.

But the prevalence of bruise in potato foods greatly increases when potatoes are handled a little more roughly and when bruises are not discarded. Normal fries and chips are basically bruise-free but about 20% of food products made from Pandora's Potatoes may contain bruise.

Consumers won't even notice.

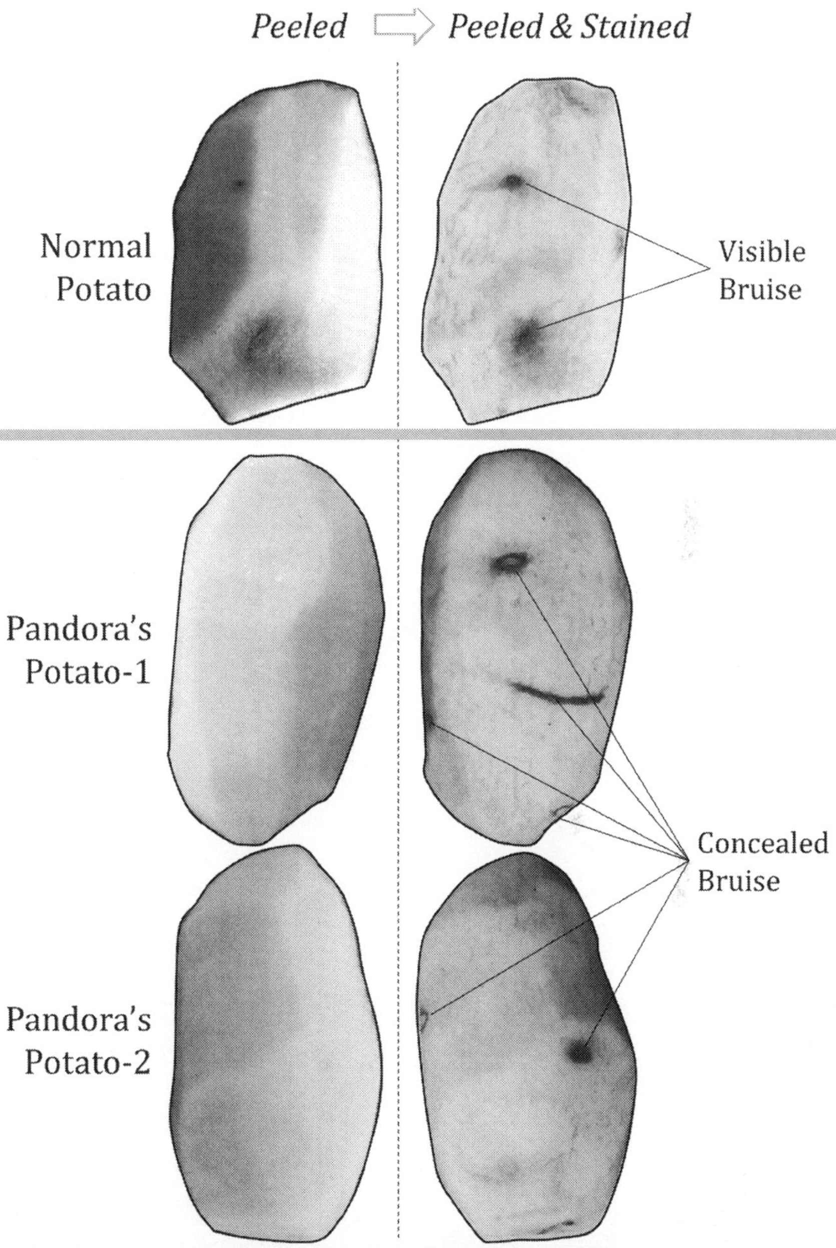

The ignored issue is that Pandora's Potatoes are **not** nutritionally equivalent to normal potatoes, especially when they contain invisible bruise. Indeed, there is nothing good or useful about the concealment of bruises, not for potatoes themseles and not for consumers.

Chaconine-Malonyl

The normal biochemistry of potatoes is disrupted by PPO-silencing, for instance, by causing the accumulation of toxins that are not present (or not as much) in normal potatoes. At least two of these toxins are formed in the entire tuber, and at least one other toxin is produced in the invisible bruises.

A potential toxin that accumulates in PPO-silenced potatoes is **chaconine-malonyl**. The levels of this glycoalkaloid-derivative are almost two-fold higher than in normal potatoes [6].

Hardly anything is known about the activities of chaconine-malonyl, but the related compound chaconine is a bitter-tasting chemical that disrupts membranes, causing symptoms including headaches, nausea, fatigue, vomiting, abdominal pain, and diarrhea.

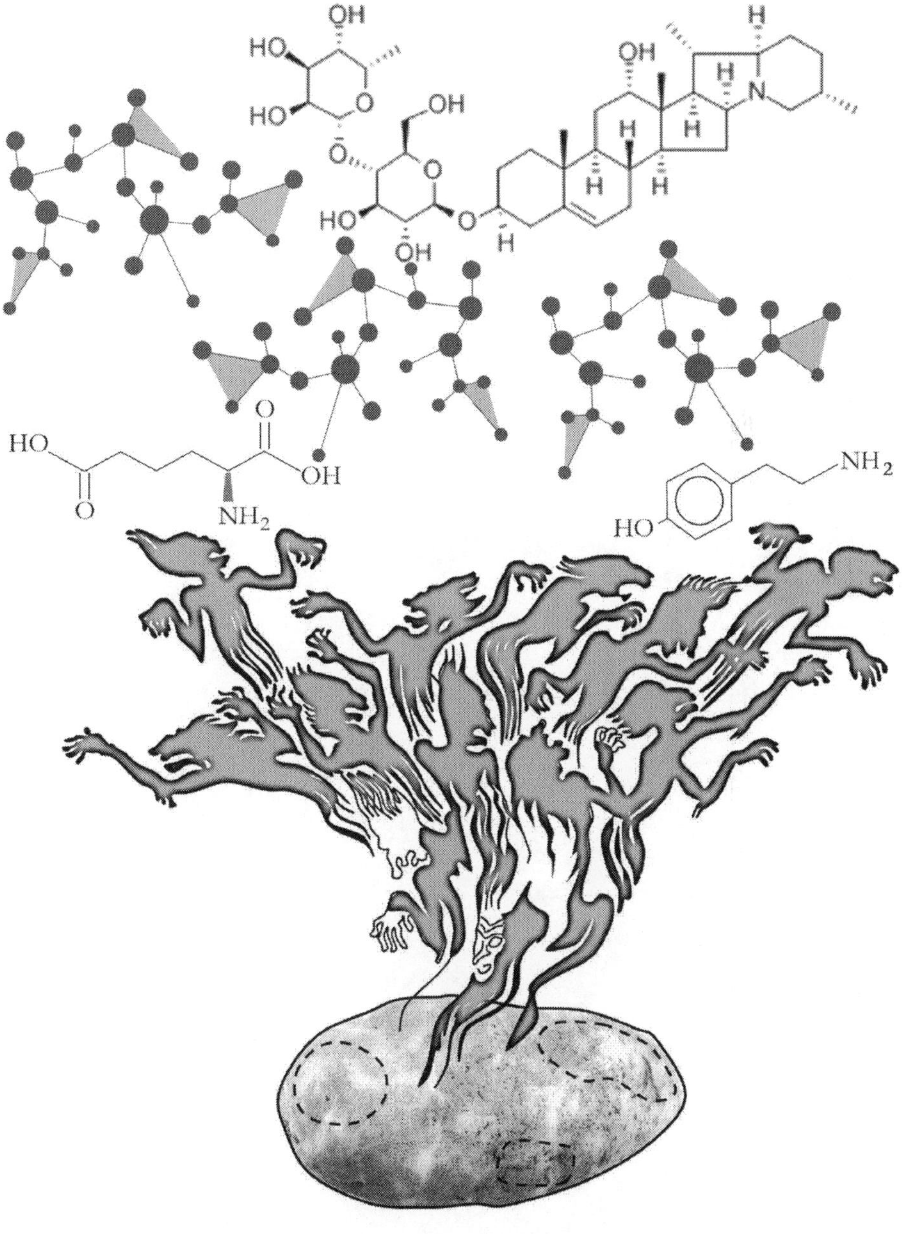

Pandora's Potatoes
RELEASED

Alpha-Aminoadipate

A second and more important toxin that is formed in PPO-silenced tubers is the lysine-derivative **alpha-aminoadipate**.

Alpha-aminoadipate is neurotoxic, affecting the astrocytes that surround neurons [7], and the compound also reacts with reducing sugars to produce a spectrum of potential hazards, including advanced glycoxidation end products. These AGEs are implicated in a variety of diet-related diseases such as diabetes, Alzheimer's, and cancer [8-12].

The level of this faulty amino acid are increased 5.9-fold upon PPO-silencing [13]. There are no published data on the alpha-aminoadipate content of Pandora's Potatoes, but normal potatoes contain about 90 mg/kg [14], which means that the modified levels may reach 90 x 5.9 = 531 mg/kg, which is almost twice as high as that of garden bean sprouts, the food source with the highest known level of this toxin.

Several years ago, concerns were raised about the high levels of dietary aminoadipate in garden bean sprouts (283 mg/kg [15]), but this issue dwarfs the potential health risks associated with Pandora's Potatoes, not only because these potatoes may contain twice as much aminoadipate per gram but also because the average consumption of potatoes (111 g/day [16]) is much greater than that of garden pea sprouts (insignificant).

The GMO corn variety LY038 was withdrawn from the European market, in part, because kernels were found to contain an elevated level of aminoadipate (56.6 mg/kg) [17-18]. This level is only one-eighth the potential level in Pandora's Potatoes.

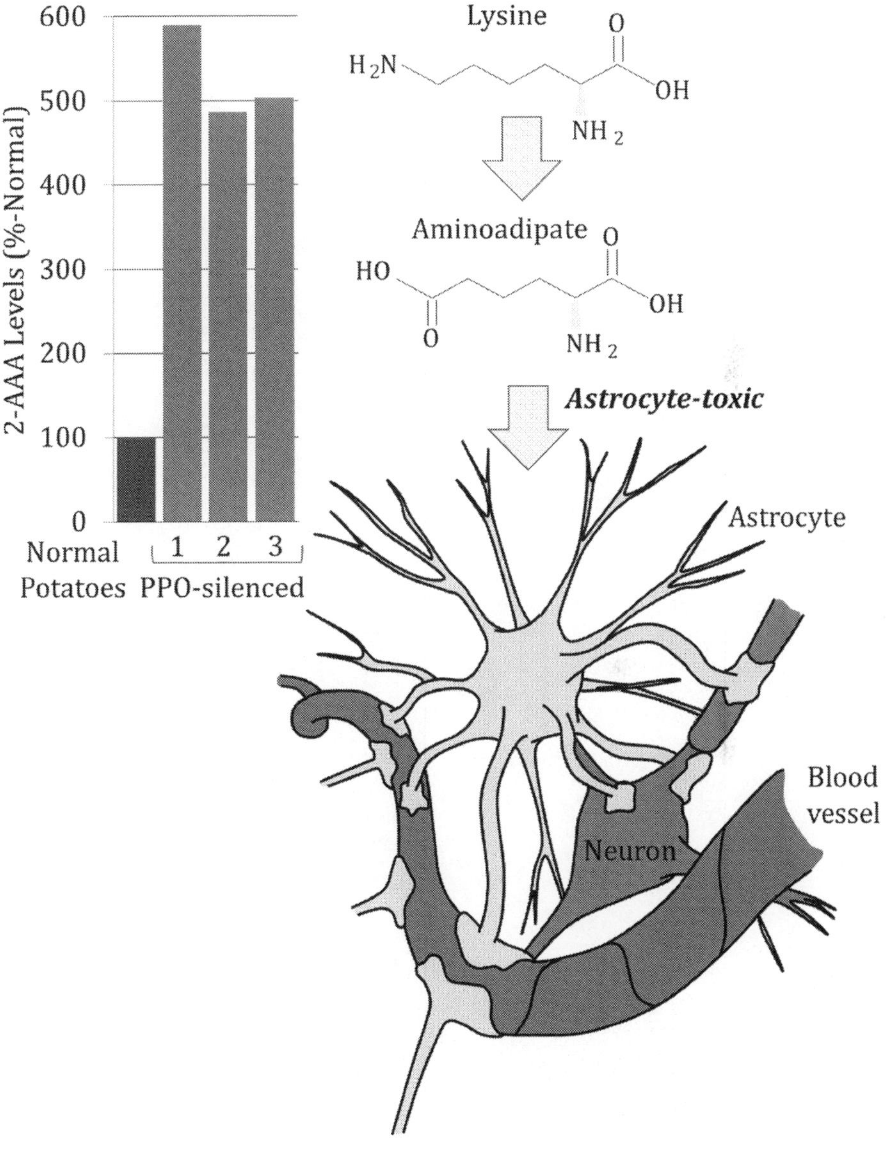
600
500
400
300
200
100
0
2-AAA Levels (%-Normal)
Normal
Potatoes
1 2 3
PPO-silenced
Lysine
H_2N
O
OH
NH_2
Aminoadipate
HO
O
O
OH
NH_2
Astrocyte-toxic
Astrocyte
Blood
vessel
Neuron

Tyramine

Another issue regarding Pandora's Potatoes is the likely accumulation of tyramine inside the bruises*.

Tyramine is, like alpha-aminoadipate, an abnormal amino acid that doesn't get incorporated into proteins. It accumulates in fermenting and rotting tissues and is a sign of death. In foods, tyramine is associated with what we call the acquired taste. It accumulates, for instance, in blue cheese—a molded milk product that, because of its smell and taste, is disliked by children but a favorite among the elderly.

Consumers may unknowingly be exposed to tyramine when they eat GMO fries, chips, or baked potatoes containing hidden and untrimmed bruises.

Normal unbruised potatoes contain trace amounts of tyramine (about 0.002 mg/g tyramine [**19**]) but tyramine accumulates to about one-fourth of all phenylpropanoids in bruised tissues** [**20,21**]. The levels of phenylpropanoids are stress-dependent and may be higher than 0.8 mg/g [**22**].

Small tubers grown in growth chambers or greenhouses will still have undetectable amounts of tyramine, but levels are expected to rise quickly in some bruised potatoes, especially when associated with specific forms of stress.

The shift from a modified Russet Burbank to a modified Ranger Russet (an alternative variety that is more susceptible to bruise) may result in a further increase in tyramine levels.

---

* There are different types of bruise that discolor in normal potatoes but are at least partially concealed in Pandora's Potatoes.

** Especially when at the cut surfaces of tuber.s

Tyramine-hypersensitive patients may develop hypertensive crisis after consumption of concealed bruise/ high tyramine Pandora's Potatoes

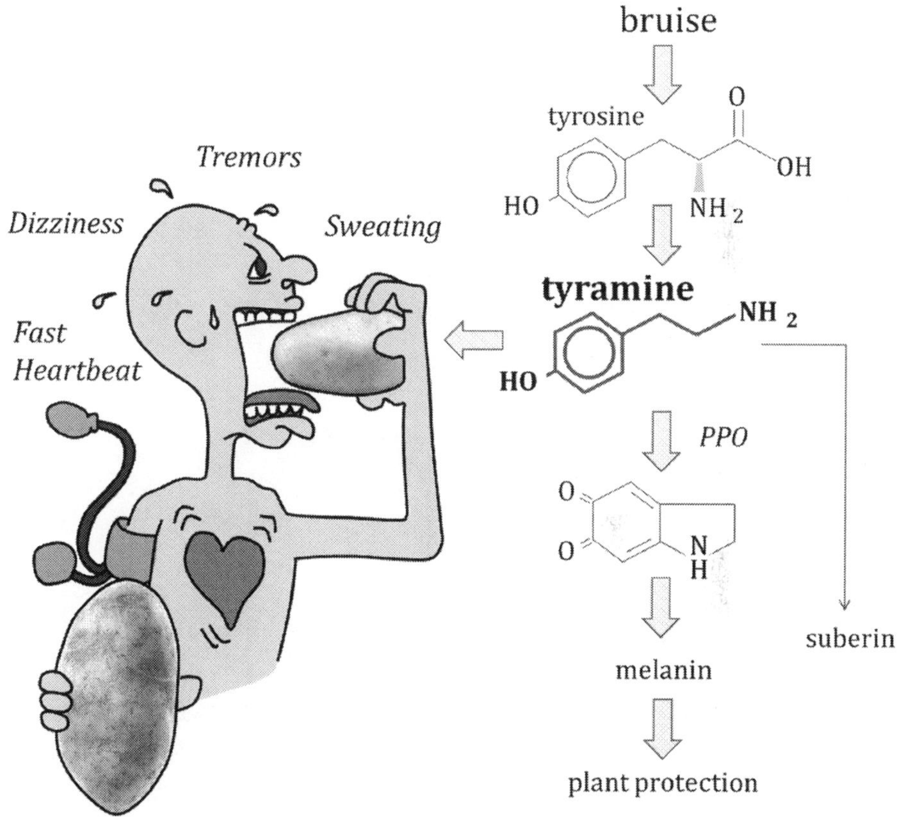

Individuals unable to metabolize tyramine (including people using MAOI antidepressants) may experience serious adverse health effects when consuming Pandora's Potatoes (baked or fried) with high levels of concealed bruise. Any ingested tyramine will accumulate in the blood and trigger the release of norepinephrine. This hormone constricts blood vessels and causes a rise in blood pressure, sometimes to dangerously high levels. Patients consuming 6-8 mg tyramine may develop elevated blood pressure, nausea, vomiting, and quickened heart rate; 10-25 mg tyramine can result in severe headache and possible bleeding in the brain (stroke), and more than 25 mg may give rise to seizure, chest pain, nausea, stroke, and hypertensive crisis (with possible complications including hemorrhage, cardiac arrhythmias, cardiac failure, pulmonary edema, and death [23]).

Tyramine-sensitive individuals avoid foods that contain high levels of tyramine, which are mostly foods exposed to extensive microbial fermentation, such as blue cheese, aged chicken livers, and soya sauce, and so the number of cases of tyramine-overexposure reported to poison control centers is limited to a few hundred per year [24]. However, there is no awareness about the potential risk associated with invisibly-bruised Pandora's Potatoes. It is possible, therefore, that unsuspecting tyramine-sensitive consumers eat GMO fries or a GMO baked potato, possibly in combination with other other tyramine-rich foods, and that the resulting spike in tyramine levels will send them to the emergency room.

Other Issues

Further research is likely to reveal additional inadvertent effects of PPO-silencing [25]. For instance, since PPO is used in normal potatoes to sequester large amounts of copper (Cu), Pandora's Potatoes will have upregulated the expression of alternative copper proteins, and some of these new proteins may represent Ra3-like allergens [26].

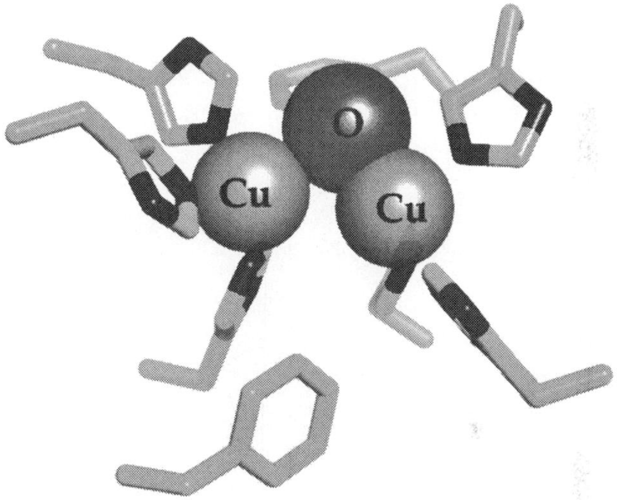

Furthermore, it is well-known that PPO-silencing affects redox-sensitive pathways in plastids [27]. This biochemical change may cause the degradation of normal amino acids into their toxic non-protein derivatives.

Given the fact that PPO-silencing has already been linked to the partial conversion of lysine into alpha-aminoadipate (and, possibly tyrosine into tyramine) [13,28], it would be sensible to test Pandora's Potatoes for elevated levels of other equally toxic amino acid derivatives.

Concealing Disease

A potato responds to infection just like how it responds to bruise: it deposits **melanin**.

But a PPO-silenced tuber cannot produce melanin, and GMO tubers will hardly develop any symptoms as they get infected by fungal mycelium, unicellular pathogens, and viruses.

For instance, a study published in 2006 shows that PPO-silencing limits late blight symptoms without limiting fungal growth and sporulation [29]:

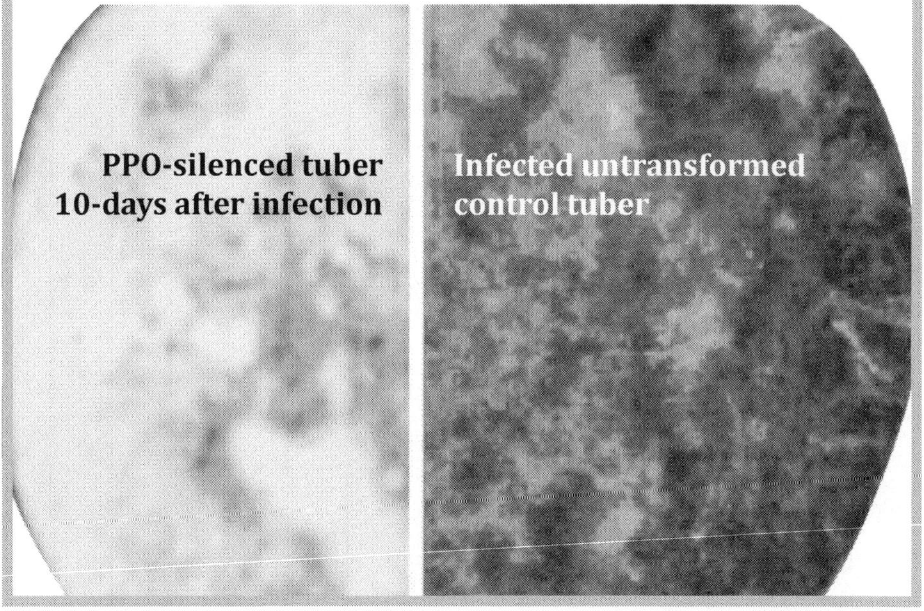

So far, it has been shown that PPO-silencing conceals the symptoms of not just late blight but also Spongospora (powdery scab), Rhizoctonia (black scurf), Fusarium (dry rot), Verticillium (wilt), Alternaria (early blight), Polyscytalum (skin spot), Sclerotinia (white mold), Rosellinia (black rot), Helicobasidium (violet root rot), Macrophomina (charcoal rot), and Helminthosporium (silver scurf) [30].

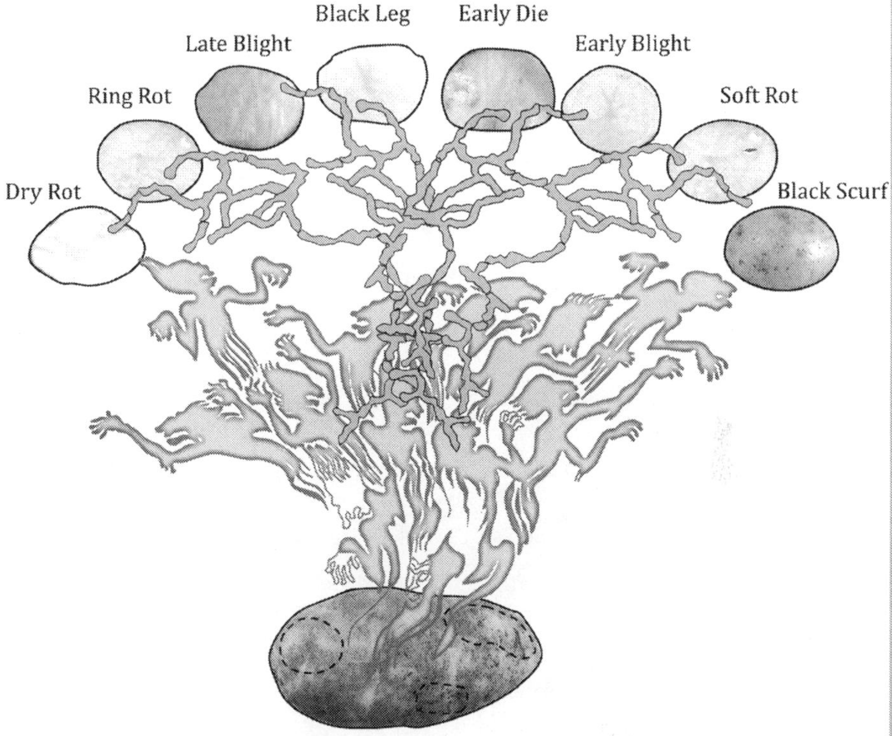

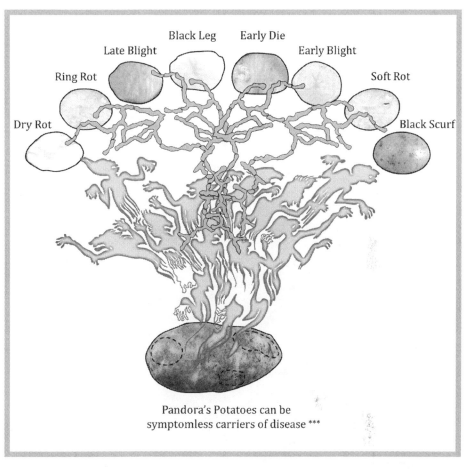

Pandora's Potatoes can be symptomless carriers of disease ***

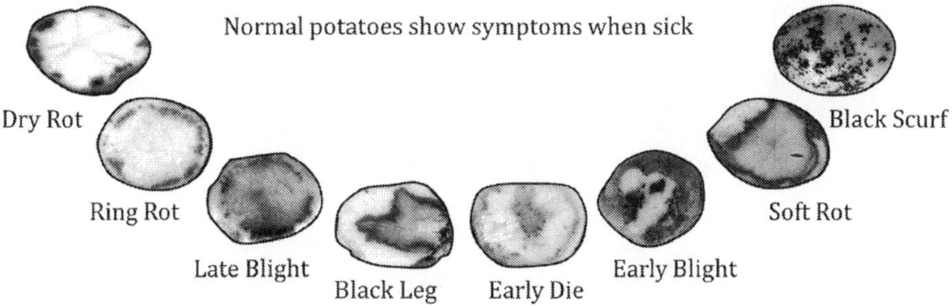

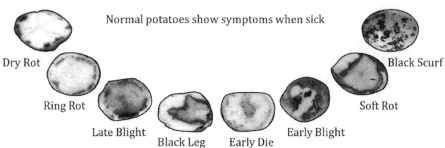

Normal potatoes show symptoms when sick

*** *Illustration is hypothetical since actual photographs are unavailable.*

Pandora's Potatoes remain mostly asymptomatic when infected. Perceived as healthy, they are shipped and mixed with other potatoes. So, billions of spores, virulent cells, and viral particles may be released even before the infection manifests itself as rot. By quietly contaminating machinery and healthy stocks, Pandora's Potatoes complicate efforts to identify disease outbreaks, contain diseases, and prevent diseases from turning into epidemics.

To avoid possible disasters, Pandora's Potatoes will have to be constantly inspected for hidden infections. Such inspections cannot be based on the simple screens for melanin deposition that are used for normal potatoes. In contrast, health certifications will need to include multiple, sophisticated laboratory tests to exclude the presence of fungal mycelium, bacterial colony-forming cells, viral particles, and so on.

Pandora's Potatoes have, so far, **not been adequately tested** for disease. Therefore, claims on tuber resistance against late blight and common scab (Streptomyces) [31] are unsupported by evidence and probably incorrect.

Because infections are **concealed**, there is a significant chance for unsuspecting consumers to be exposed to the toxins, carcinogens, and allergens produced by bacterial and fungal pathogens. Some of the hundreds of **toxins** that may accumulate in Pandora's Potatoes include:

> *AAL-toxin,*
>
> *tentoxin,*
>
> *zinniol,*
>
> *lycomarasmin,*
>
> *Rhizoctonia toxin,*
>
> *Pectobacterium toxIN ribonuclease,*
>
> *coronatine,* and
>
> *Verticillium toxin.*

Pandora's Potatoes may also be infected with Fusarium or other pathogens producing **carcinogens** such as *Fumonisin*.

Furthermore, fungal pathogens such as Alternaria, Fusarium, Verticillium, and many others produce potent **allergens** including *alt a 1* that may trigger serious allergic reactions in sensitive consumers.

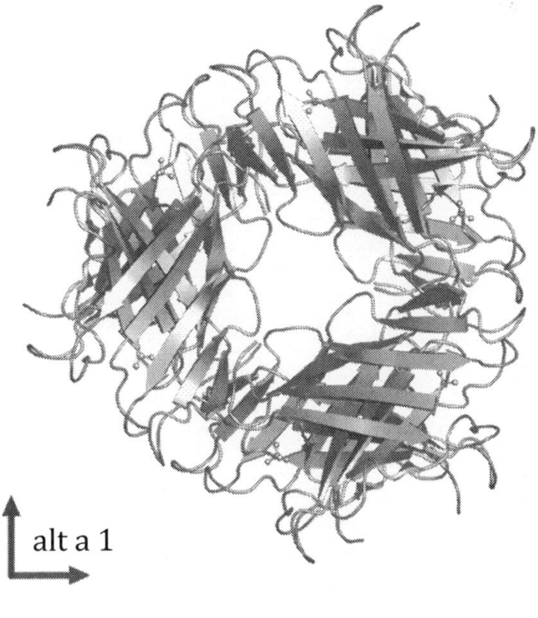

Impaired Melanin Formation

Melanin is a protective, non-digestible, dark layer of oxidized and polymerized phenolic compounds produced by not just potatoes but also bananas, avocados, apples, and numerous other plants, animals, fungi, and bacteria. All these living things have preserved PPO for millions of years because they need melanin for long-term survival.

In potato tubers, melanin quickly accumulates when healthy tissues are bruised and structurally weakened. The melanin is intended to protect the damaged tissues, which have become potential **entry points** for infection; it slows down the rate of infection and infestation by fortifying and protecting the bruises, like steel fortifies concrete.

Equally important is dehydration. Bruises represent **exit points** for water loss, but this undesirable process is somewhat diminished through melanin deposition.

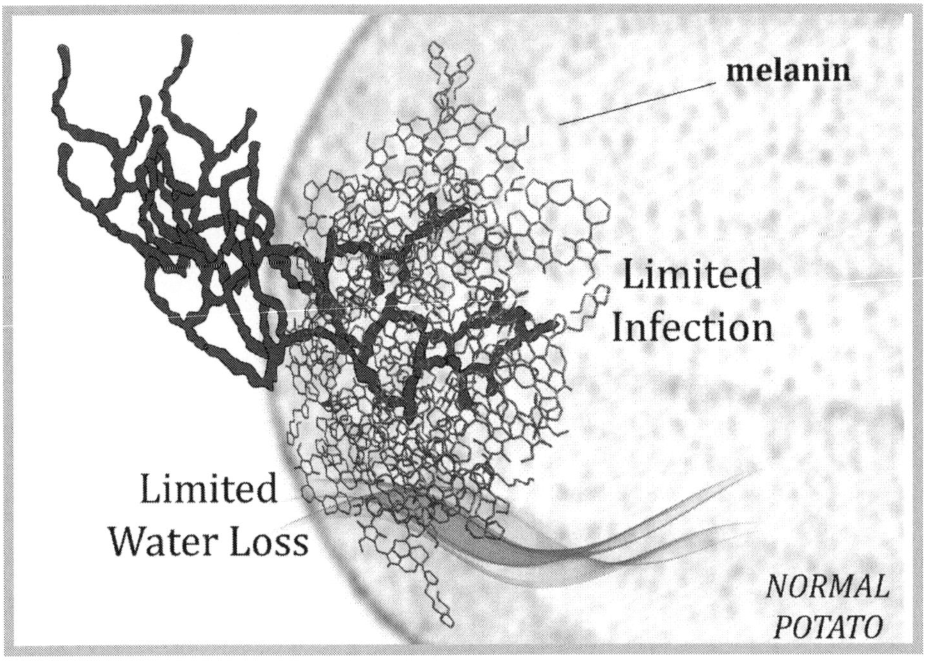

Healthy tubers also deposit melanin wherever unbruised tissues are compromised by infection, infestation, or toxins. Melanin deposition is part of the basic plant stress tolerance response, as has been confirmed by thousands of scientific studies. The ancient response provides partial protection against abiotic stresses (such as high temperatures, intense UV radiation, dehydration, and toxins) as well as numerous biotic stresses (virulent microbes and pests) [32].

More specifically, melanin stress-tolerance has been shown to play a role in the control of, for instance, soft rot [33], Colorado beetles [34], bacterial wilt [35], bacterial leaf spot [36], bacterial speck [37], Pseudomonas, cutworm, bollworm, beet army worm [38], forest tent caterpillars [39], downy mildew [40], Fusarium [41], oxidative stress [42], water stress [43], drought stress, and salt stress [44].

Based on a recent literature analysis, I believe that potatoes unable to deposit melanin are at least 5% more vulnerable to rot, dehydration, and stress than normal potatoes. The resulting yield losses may reduce profits by 50%, from about $300 to $150 per acre.

1

Melanin, the Indispensable Marker

Melanin deposition **marks** potatoes that are compromised in quality and must be discarded. The marker may seem wasteful to processors but it **prevents an escalation of agronomic issues** on the farm. Since Pandora's Potatoes lack this internal quality control system, farmers lost their ability to monitor for disease and bruise; what used to be manageable issues turned into silent threats that may destroy livelihoods.

Without the melanin marker, it is much more difficult to contain diseases. Infected potatoes that seem healthy may be transported to other locations and cause pathogens to spread in ways that were inconceivable in the past.

Furthermore, the inability of Pandora's Potatoes to darken their bruises may result in a new attitude among farmers. Unable to rely on a marker for damage, they may become just a little less careful in avoiding damage; they may treat Pandora's Potatoes a little more roughly during harvest and transport than they would have treated normal potatoes. It would not be surprising if, during the subsequent storage, the incidence of concealed bruise in Pandora's Potatoes would be much higher than the incidence of visible bruise in normal potatoes, especially because there won't be any inspections for certain types of bruise. Thus, pathogens are likely to find more entry points for infection, and none of these entry points is protected by melanin. I expect that the incidence of post-harvest diseases such as Fusarium dry rot, late blight, early blight, bacterial soft rot, and Pythium leak will increase [**45**,**46**], and that the associated losses will double from $100 to $200 per acre [**47**].

Additionally, the water that is stored inside potatoes will escape through many more exit points, and these exit points are now void of protective melanin [**47**]. Any increase in water loss will exarcebate the important 'shrink' (yield loss) issue. A farmer who would normally lose $100 per acre on shrink may have to deal with much greater losses, perhaps $200 per acre, after storing Pandora's Potatoes.

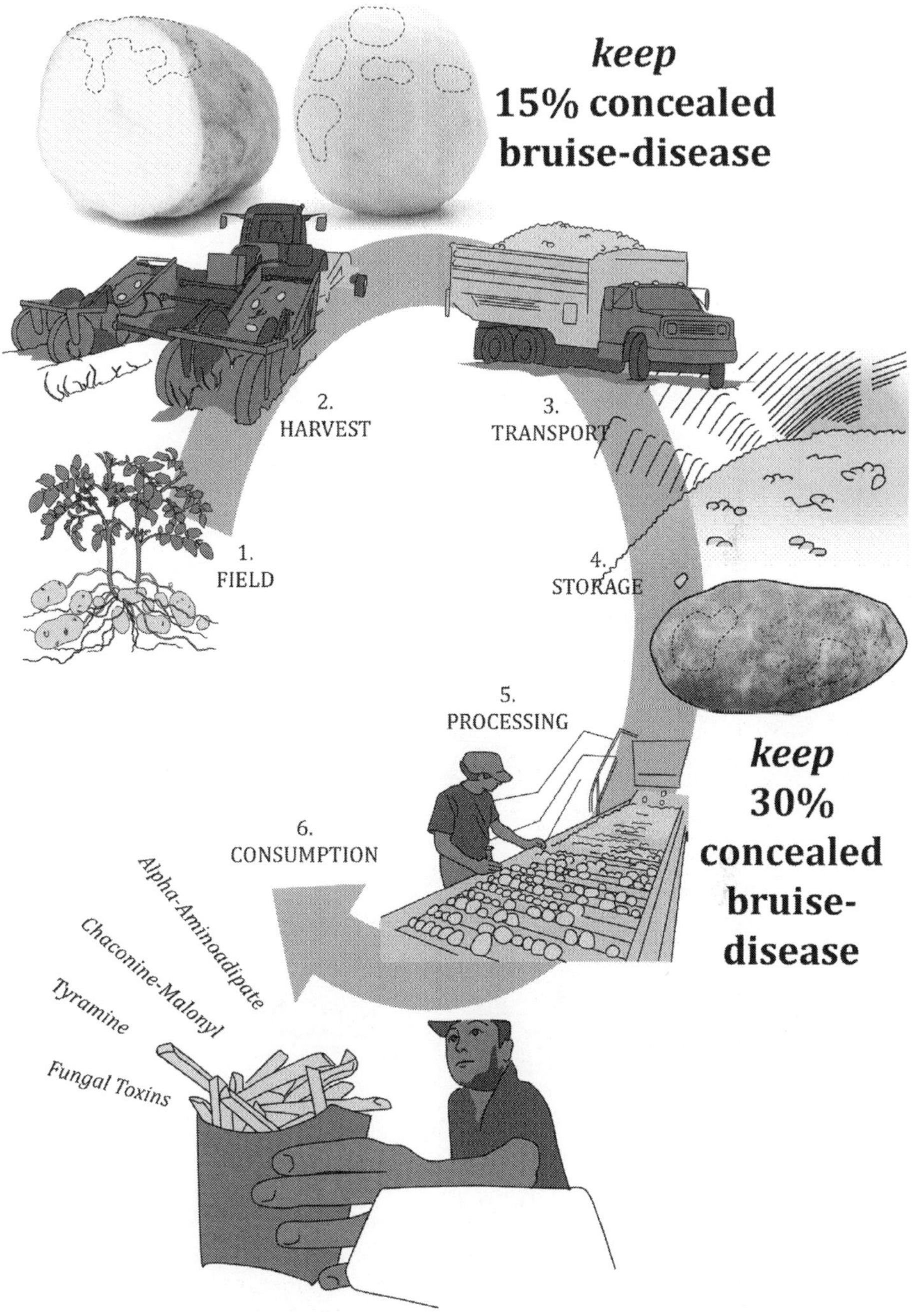

The melanin marker of normal potatoes also **keeps toxins from the plate**. Without this marker, consumers are now exposed to the known and unknown toxins, carcinogens, and allergens accumulating inside bruised and infected potatoes.

2

ASN-SILENCING

Permission to Eat

MOST PEOPLE LIMIT their consumption of French fries because they know that the intake of large amounts of fat, salt, and calories causes health issues.

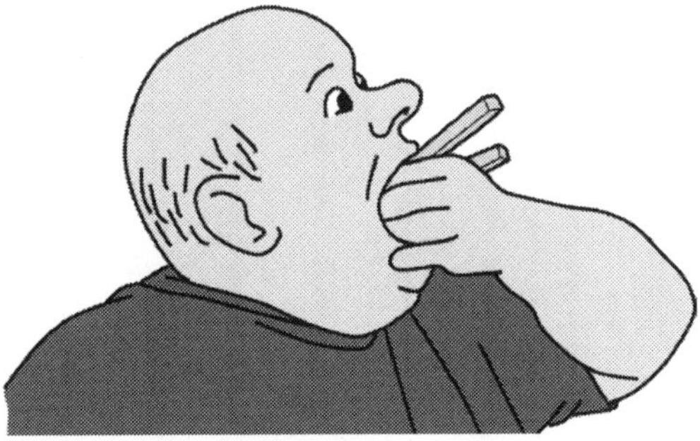

But they may be convinced to eat more fries when given an excuse, when they are told that fries are healthier now.

This strategy was already used in the 1960s to increase the sales of cigarettes: people were told that low-tar cigarettes are healthier than normal cigarettes. It was a lie that worked.

Consumers are now taught that French fries are healthier because they contain less acrylamide. Acrylamide is a natural compound that is carcinogenic only at levels 1,000-10,000 times higher than the levels in fries [48]. In other words, lowering the acrylamide levels in French fries is lowering the insignificantly-low levels of a carcinogen to even more insignificantly-low levels.

The advertisement on the left is from 1994 and similar to many other advertisements that are now disallowed by the FDA. The advertisement on the right is hypothetical and represents the current strategy of potato processors.

Evidence published in 1996 demonstrated that the mutagenicity (indicator for carcinogenicity) of fried potatoes is 1,000- times lower than the mutagenicity of fried beef [49].

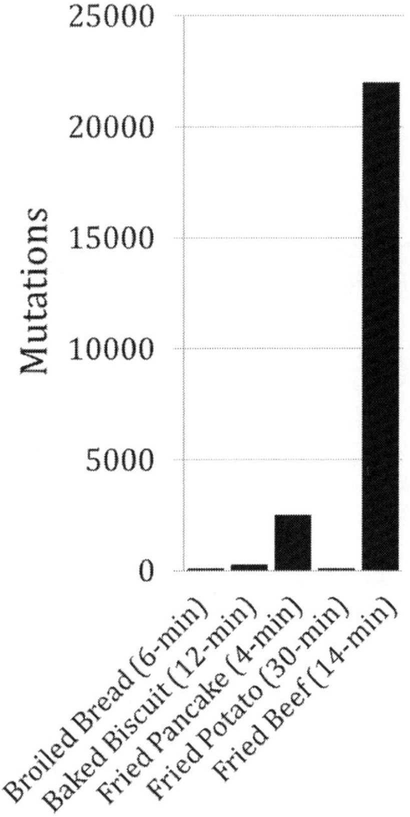

Furthermore, epidemiological studies have shown that dietary acrylamide is not linked to cancer [50,51]. An objective person knows that it is as impossible to get cancer from eating fries as it is to drown from drinking a glass of water.

It doesn't seem to matter that the health claim is fake. A fake health claim increases the perceived quality of a food just as much as a true health claim—most of the food experience is about perception.

Wasting Nitrogen

The ASN gene plays an important role in the assimilation, storage, and use of **nitrogen**, which is the most important building block of proteins [52]. Therefore, silencing of ASN is likely to compromise the plant's ability to assimilate nitrogen in roots, which means that Pandora's Potatoes are expected to be less efficient in the use of nitrogen-fertilizers. Even a 5% reduction in nitrogen-use efficiency costs farmers $10 per acre.

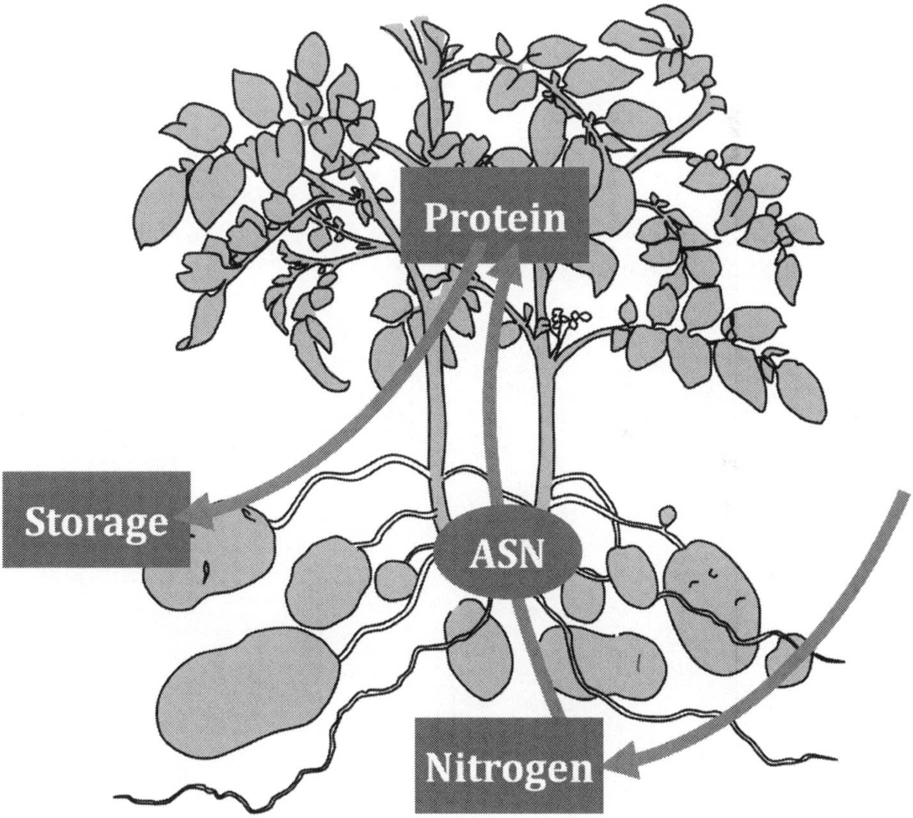

INV-SILENCING

CONSUMERS MAY NOT NOTICE the effects of PPO- and ASN-silencing but they will certainly miss INV, which is the gene responsible for almost all the sensory attributes of fried, roasted, and baked potatoes. Double-transformed versions of Pandora's Potatoes lack the INV-produced glucose and fructose that, when heated, get converted into the typical golden-yellow colors, attractive aromas, and desirable flavors of normal potato foods.

We eat fries with our eyes because we know intuitively that the intensity of color is indicative of all sensory characteristics. Indeed, taste tests have confirmed that golden-yellow fries smell and taste better than creme-colored fries [53,54]. This means that, compared to normal fries, GMO fries have an inferior sensory profile.

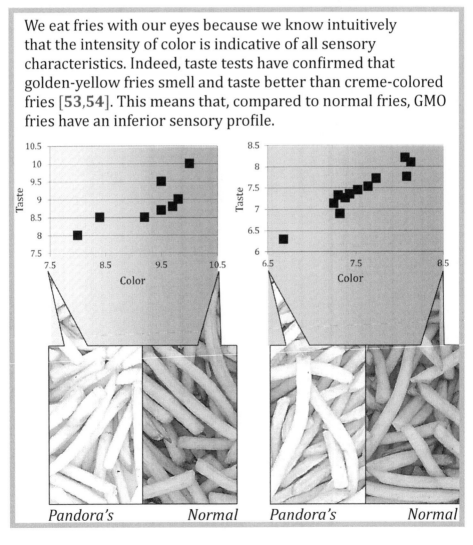

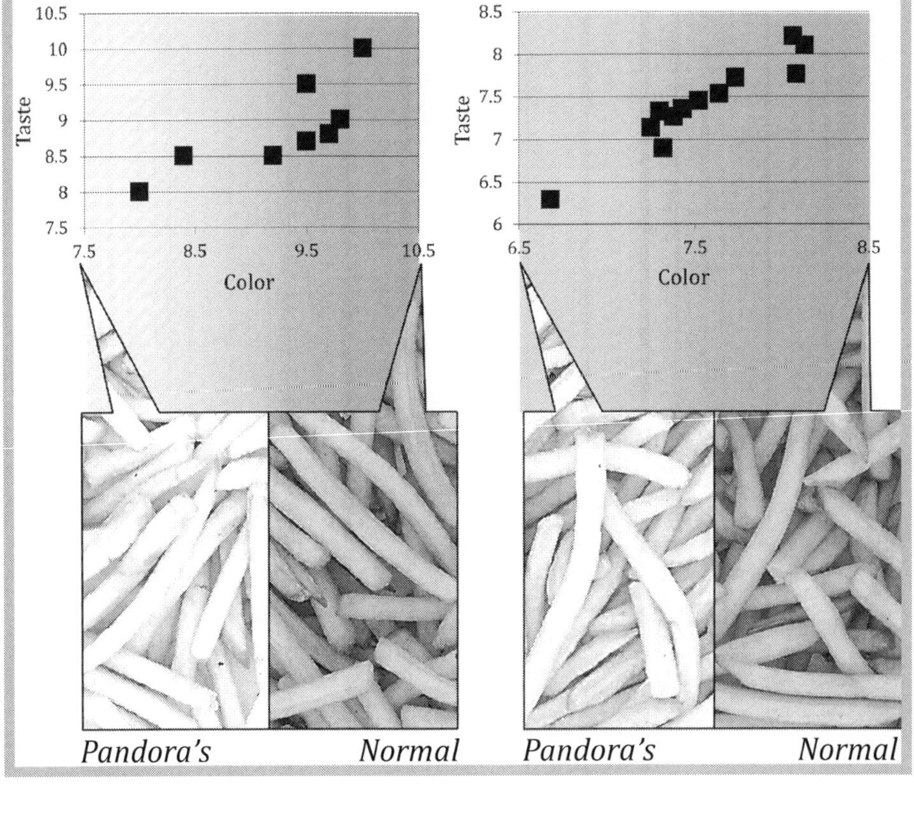

Processors are aware of INV's critical role in food quality but they also know that INV causes product diversity. For instance, the ends of fries may be darker then the central parts, and fries from fresh potatoes are often lighter than fries from stored material.

This product diversity is hard to accept for people working in the food industry. These people want their products to be always exactly the same. They rather produce uniform GMO fries that look and taste like cardboard than the normal fries that consumers prefer.

Apparently, consumers are expected to pay the price of processor convenience.

Agronomic Issues

The soluble sugars glucose and fructose that cannot be produced anymore in the tubers, roots, stolons, and flowers of Pandora's Potatoes are not just important for food quality but also for physiologically-important processes such osmoregulation, cryo-protection, and membrane stabilization [55-57]. Consequently, Pandora's Potatoes are delayed in emergence [30,58] and may also be compromised in fertility (as has been shown for INV-silenced cotton [59], and so on.

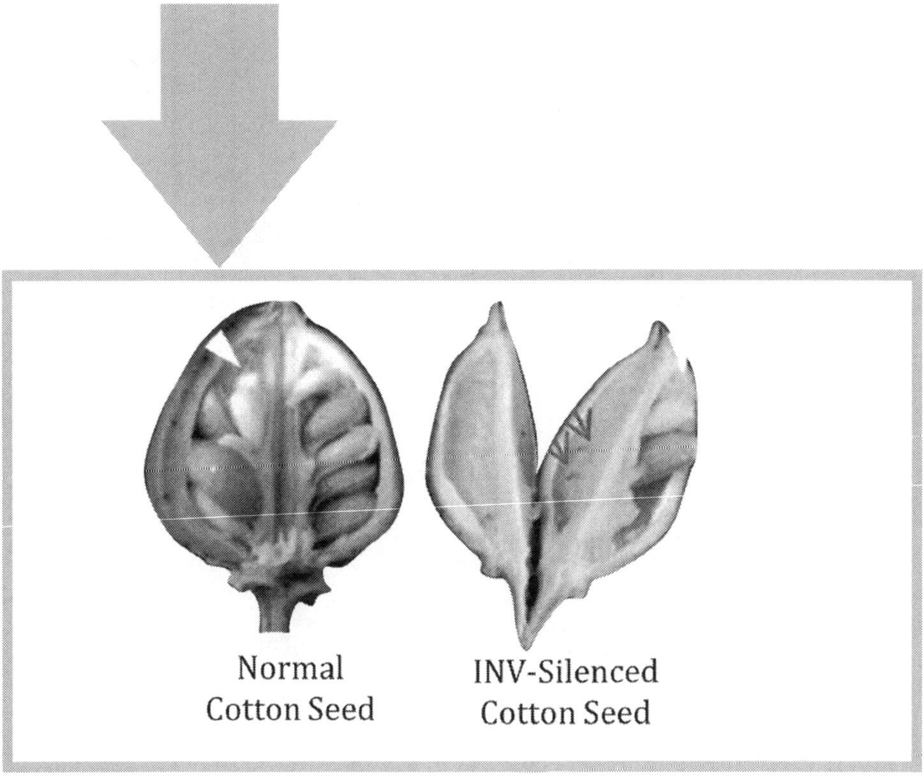

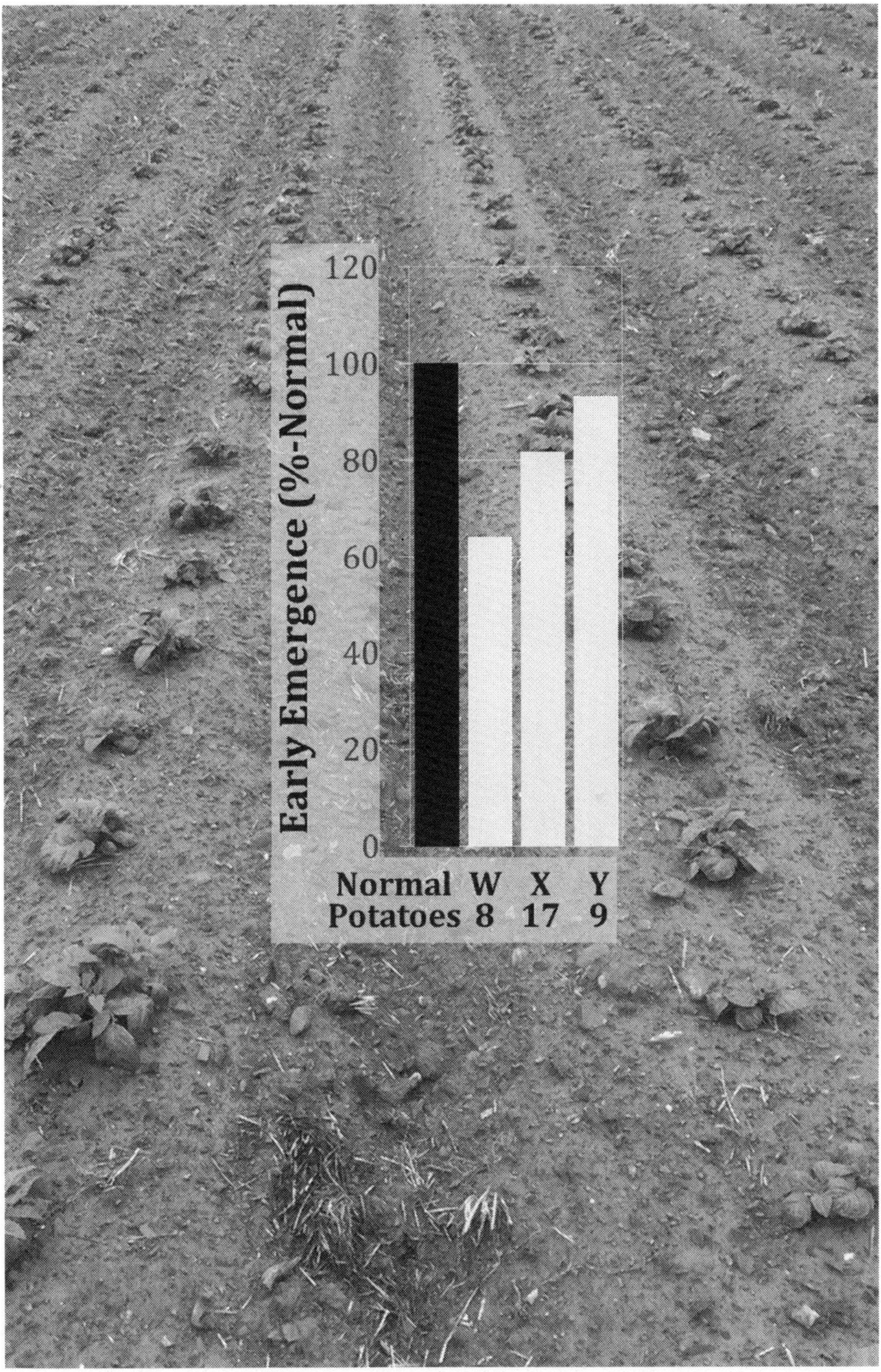
Early Emergence (%-Normal)
120
100
80
60
40
20
0
Normal Potatoes
W 8
X 17
Y 9

4

VNT-INSERTION

Biopiracy

PANDORA'S POTATOES represent the first example of a GMO crop that contains an illegally acquired exotic gene. This gene, VNT, is not derived from a freely-available domestic organism but was isolated, without authorization or compensation, from a wild plant growing in Argentina.

Article 15 of The Convention on Biological Diversity affirms a State's sovereign rights over its genetic resources, which means that **VNT belongs to Argentina** and that the illegal isolation and commercialization of this gene by foreign scientists is an act of **biopiracy**.

In the past, biopiracy was limited to chemicals, and companies such as Bristol-Myers Squibb and WR Grace made billions of dollars isolating and exploiting chemicals such as Captopril and Azadirachtin. These practices are not acceptable anymore, and the pharmaceutical and agrochemical industries have entered into benefit-sharing agreements with local governments to exploit the biodiversity of exotic plants and animals. However, genetic engineers are still in their *conquistador phase* as they steal R-genes from exotic plants—not just the Argentinian VNT gene but also the Mexican R8 and AMR3 genes, the Peruvian PHU1 gene, and many others. It took wild potatoes millions of years to develop R-genes that play important roles in local ecosystems. But the theft and global overexploitation of these genes exhausts a most-depletable resource and compromises the competitiveness of wild potatoes in their original habitat.

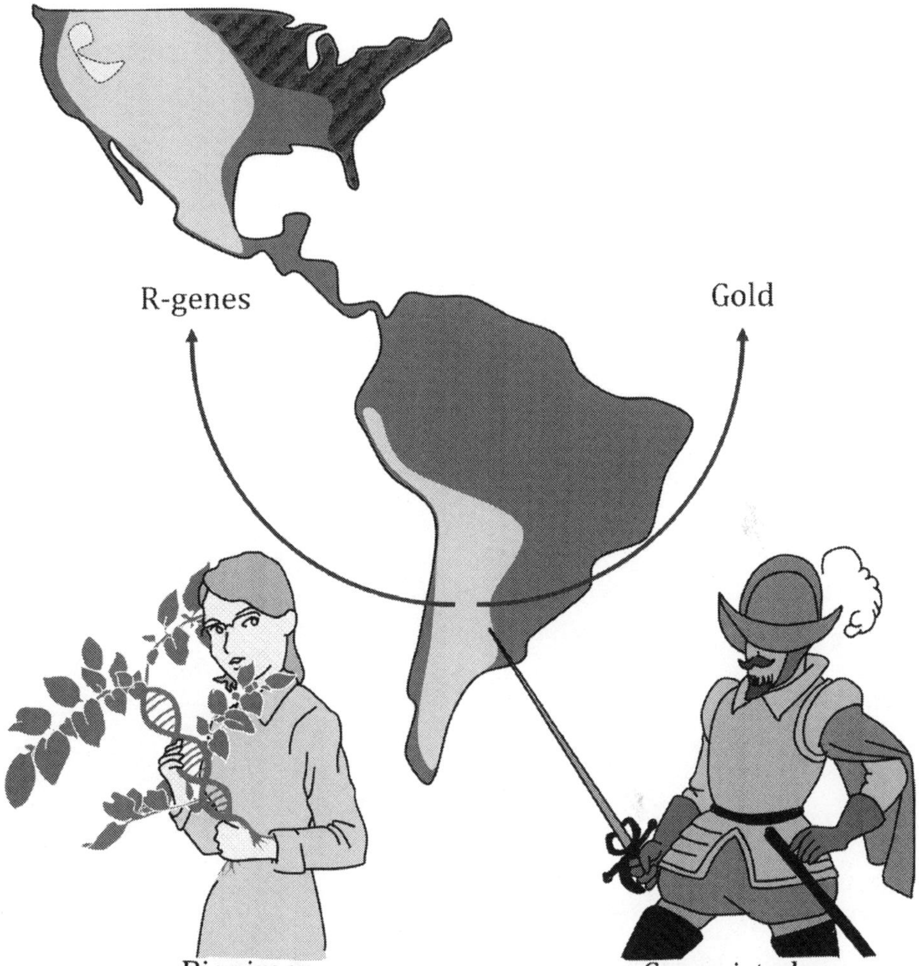

Many mistakes have been made so far, and the value of VNT is much lower than the hundreds of millions of dollars estimated by processors and their academic collaborators [**60-62**], in part because it is offset by yield drag and compromised stress tolerance (as discussed previously). Regardless of what the actual value may be, it seems reasonable to expect that profits will be shared with the Argentinian government.

4

Why Was VNT Introduced Into Potatoes?

Late blight is easily managed where most potatoes are grown in the United States, which is where potatoes do best and diseases and pests do worst (warm days, cool nights, optimal moisture for the naturally-resistant roots, and no extended leaf wetness). Two-thirds of all potatoes are harvested from farms in parts of the Snake River Plain and the Columbia Basin representing less than 0.02% of the United States landmass.

In the arid and irrigated Northwest, losses to the disease are mostly insignificant. The replacement of normal potatoes by Pandora's Potatoes limits cost savings (late blight fungicide applications) to a maximum of $50 per acre [**63**], which is only a fraction of the losses caused by yield drag and stress sensitivity.

The situation seems different for the agricultural regions where the climate facilitates mycelial infection and the spread of fungal spores. These more **humid regions** include the Northeastern part of the United States where farmers growing Pandora's Potatoes may be able to limit their late blight fungicide sprays by 50% while also preventing a 5% yield loss due to infection at a cost saving of $200 per acre [**61,62,64,65**]. It is not surprising, therefore, that the GMO varieties are already grown in *Late Blight Territory*—not just in states such as New York but also in countries as vulnerable and far away as Bangladesh and Indonesia [**66**].

Indeed, the developer of Pandora's Potatoes attempts to maximize the return on its investments by using VNT where it seems most valuable, even though this overexploitation of the gene minimizes its lifespan and hastens the need for stacking and more biopiracy.

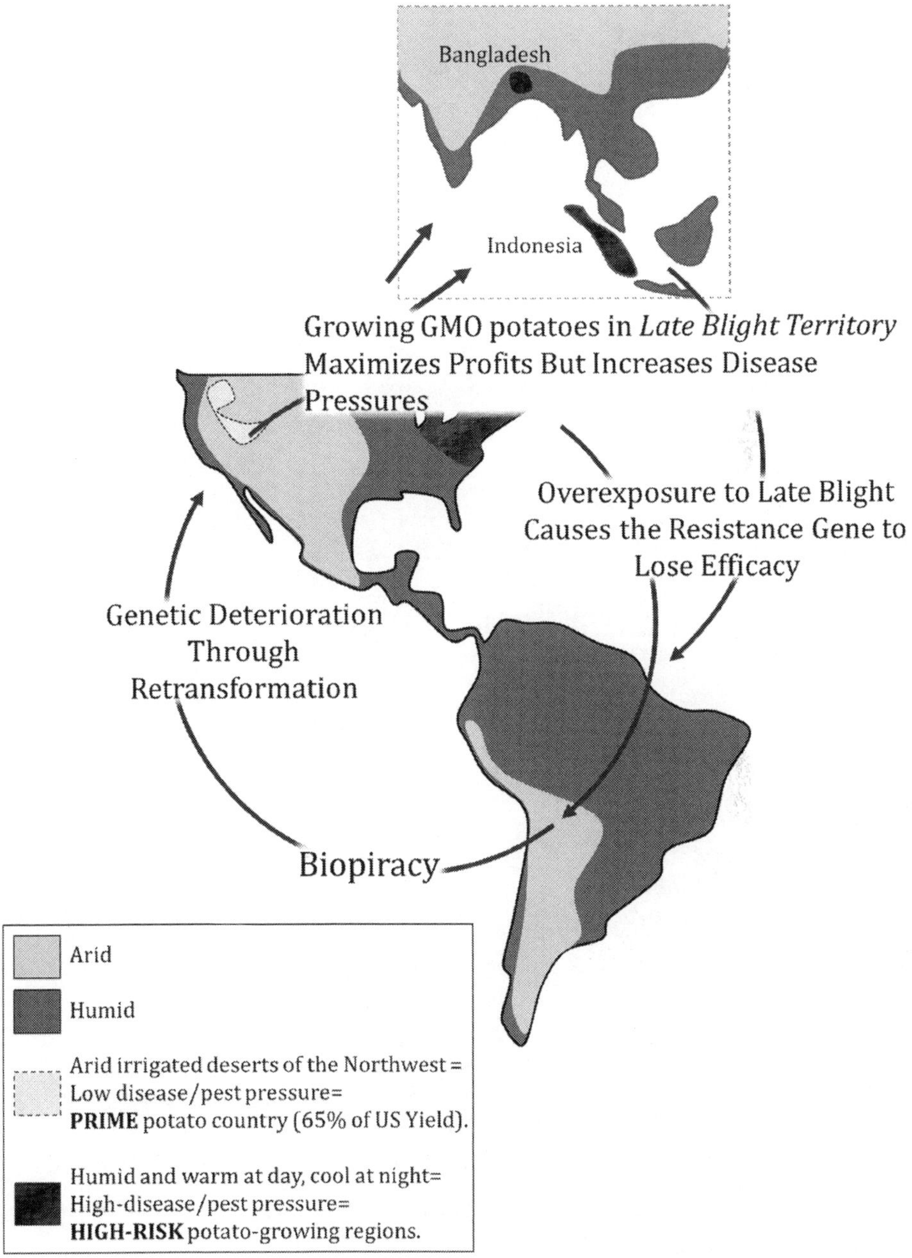
Bangladesh
Indonesia
Growing GMO potatoes in *Late Blight Territory* Maximizes Profits But Increases Disease Pressures
Overexposure to Late Blight Causes the Resistance Gene to Lose Efficacy
Genetic Deterioration Through Retransformation
Biopiracy
Arid
Humid
Arid irrigated deserts of the Northwest = Low disease/pest pressure= **PRIME** potato country (65% of US Yield).
Humid and warm at day, cool at night= High-disease/pest pressure= **HIGH-RISK** potato-growing regions.

4

Late Blight Territory

The strategy to grow Pandora's Potatoes in *Late Blight Territory* is as shortsighted as the hypothetical plan to vaccinate soldiers for a single tropical disease and then deploy them into the tropics. The soldiers would contract any of the other diseases they had not been protected against, just like Pandora's Potatoes growing in New York or Indonesia are targeted by dozens of virulent diseases and pests.

Indeed, the apparent intent to change the geographical redistribution of potato farms, from the arid Northwest to *Late Blight Territory* does not consider the fact that **late blight never comes alone!** Whenever the conditions are ideal for late blight, they are also ideal for many other diseases and pests.

In New York, the average damage caused by late blight represents only a small part of the average damage caused by other organisms [**67**]:

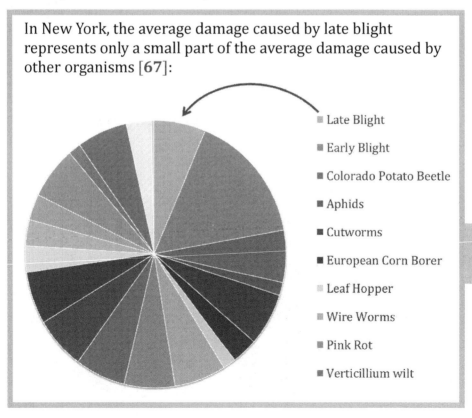

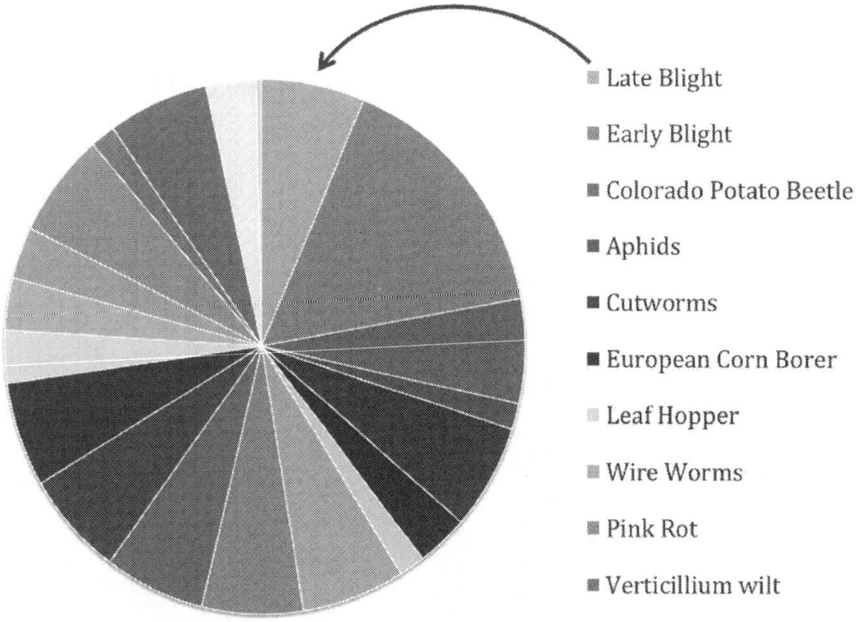

Farmers in *Late Blight Territory* have to address all diseases and pests as a single complicated issue by relying on combinations of fungicides and insecticides with the broadest-spectrum activities. It is the only way to limit protection costs to less than about $1,200 per acre.

Even though Pandora's Potatoes provide some late blight resistance, farmers growing these potatoes cannot stop applying late blight fungicides. Continued sprays are needed to control other diseases, such as the equally important early blight, and also as insurance, in case late blight virulence suddenly re-emerges (and to slow down the development of such re-emergence).

An additional issue is that any increase of the potato acreage inside *Late Blight Territory* increases the breeding grounds for diseases and pests that target potatoes. A doubling of the acreage in the Northeast, from 65,000 to 130,000 acres, is estimated to increase the per-acre costs of diseases and pests by five to ten percent [**68**], which means by $50 to $100 per acre.

Furthermore, the replacement of numerous potato varieties by a minimum of Pandora's Potato varieties reduces the genetic diversity of potatoes and so aids pathogens and pests in fine-tuning their invasiveness [**69,70**], which means that infections and infestations become more aggressive and spread more quickly.

The only sector of the industry that benefits from an increased need for fungicides, pesticides, and fertilizers is the **agrochemical business**, which includes the company developing Pandora's Potatoes.

Unstable Resistance

The resistance (R-) gene used to control late blight belongs to a large class of genes extensively tested during the 1950s to 1980s. Breeders have already transferred hundreds of R-genes from wild species to domesticated plants (11 genes for late blight control alone), but their attempts to develop durable resistance were mostly unsuccessful, and it was concluded that **three issues limit the applicability of individual R-genes**:

- An R-gene never controls a disease but only a fraction of microbial strains causing that disease;
- Even strains controlled by an R-gene have the genetic versatility to co-evolve quickly and regain their virulence because they contain numerous virulence factors, propagate sexually, and grow on dead plant tissue [71,72].
- The composition of late blight strains is naturally diverse and changes from year-to-year in unpredictable ways [73]. For instance, the profile for 2014 was completely different from the 2009 profile:

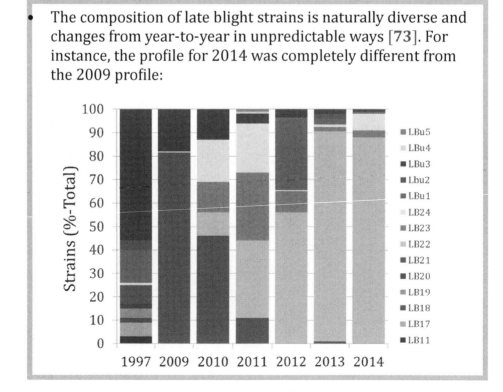

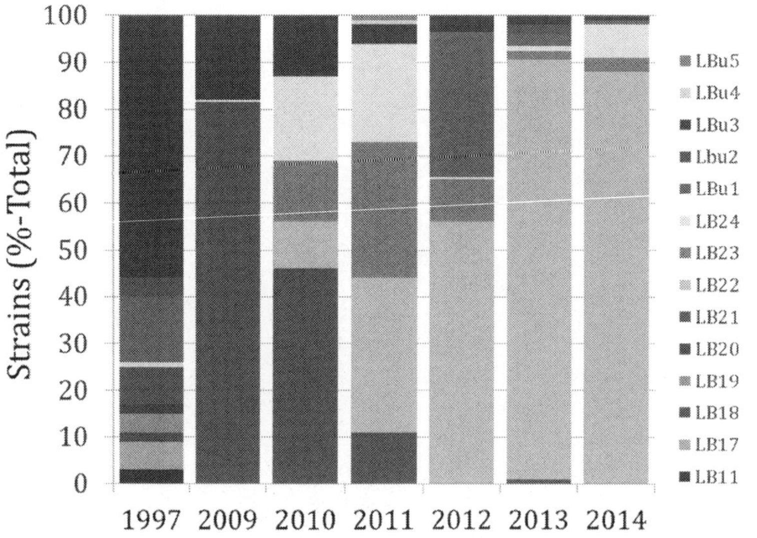

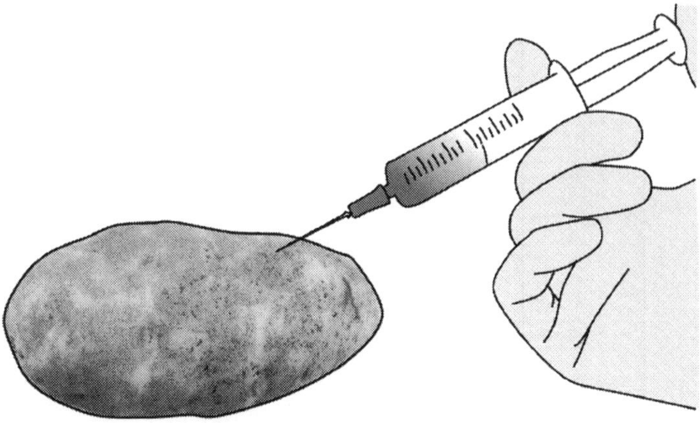

Thus, an R-gene that is functional during one year may not be functional the next. It is **like a flu shot** in that it provides resistance to some of the numerous strains causing a disease. But the R-gene is unlike the flu shot in that it cannot be updated and re-applied each year. This means that any R-gene employed may be effective for a few years at most.

Based on their disappointing experiences, most breeders have abandoned their attempts to develop R-gene based resistance and, instead, refocused their efforts on the more durable and broad-spectrum horizontal resistance provided by multiple genes [74,75].

Pretending to be smarter, genetic engineers have been promising (for more than two decades now) to succeed where breeders had failed. They will develop durable resistance by using single R-genes. Their promise is as unrealistic as the promise of a universal flu vaccine, and it is equally effective in securing millions of dollars in research funding. Indeed, many genetic engineers turned R-genes into lucrative, lifelong careers.

And so history repeats itself.

R-genes are again tested for field efficacy and hope is again turned into disappointment. One of the first transgenic R-genes tested, BLB1 (also called RB [76]), was promised to maintain its unusual broad-spectrum resistance under intense disease pressure [77]. Ironically, though, this gene appeared even less stable than the conventional R-genes.

Subsequently tested transgenic R-genes were also defeated before commercialization [78,79], and genetic engineers eventually came up with VNT [80,81], the gene that was introduced into Pandora's Potatoes.

VNT has not yet been broadly exposed to late blight but its weaknesses have already been revealed:

- European and Central American strains including EC1 and NL11479 already evolved around VNT [82].
- Even strains controlled by the R-gene become virulent when infections occur during the mid or late growing season [83]:

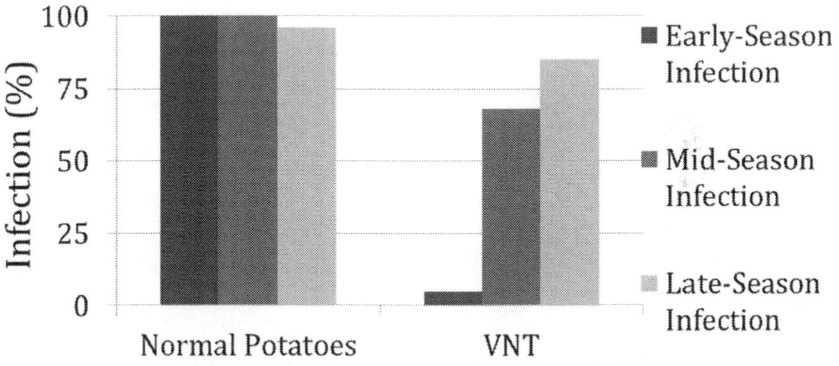

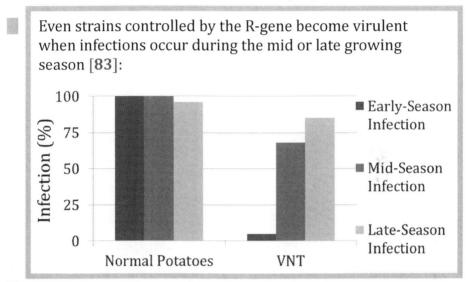

- Infection was delayed but not prevented during a field trial in Europe [79].
- Even the controlled strains are partially virulent in tubers [58].
- The gene is often not expressed or functional in tubers [58].

Importantly, these issues became apparent after simple laboratory tests and small field trials. The real-life situation, when thousands of acres of Pandora's Potatoes are exposed to aggressive late blight strains, will prove to be much harsher.

The resistance mediated by R-genes such as VNT can never be guaranteed. What can be guaranteed is that the resistance will be broken, at any time, by new or mutated strains of the disease. It is common knowledge among plant pathologists that *the more* a disease is confronted with an R-gene, *the faster* it evolves around this barrier to infection. Farmers growing Pandora's Potatoes in *Late Blight Territory* will soon be confronted with new, virulent strains and a re-emergence of the disease.

It is possible to prolong the resistance by adding new R-genes with different specificities, such as the Mexican R8 and AMR3 genes, or the Peruvian PHU1 gene. However, gene stacking efforts take at least eight years to complete (transformation, field testing, regulatory approval, bulk-up, and commercialization), which is too slow to respond to the unpredictable behavior of the late blight fungus. And there is also an agronomic cost to stacking: each re-transformation of Pandora's Potatoes causes the accumulation of hundreds of new mutations, resulting in the additional loss of at least five-percent yield potential.

The bottom line is that late blight always wins. The pathogen is much faster and smarter than the fastest and smartest genetic engineer.

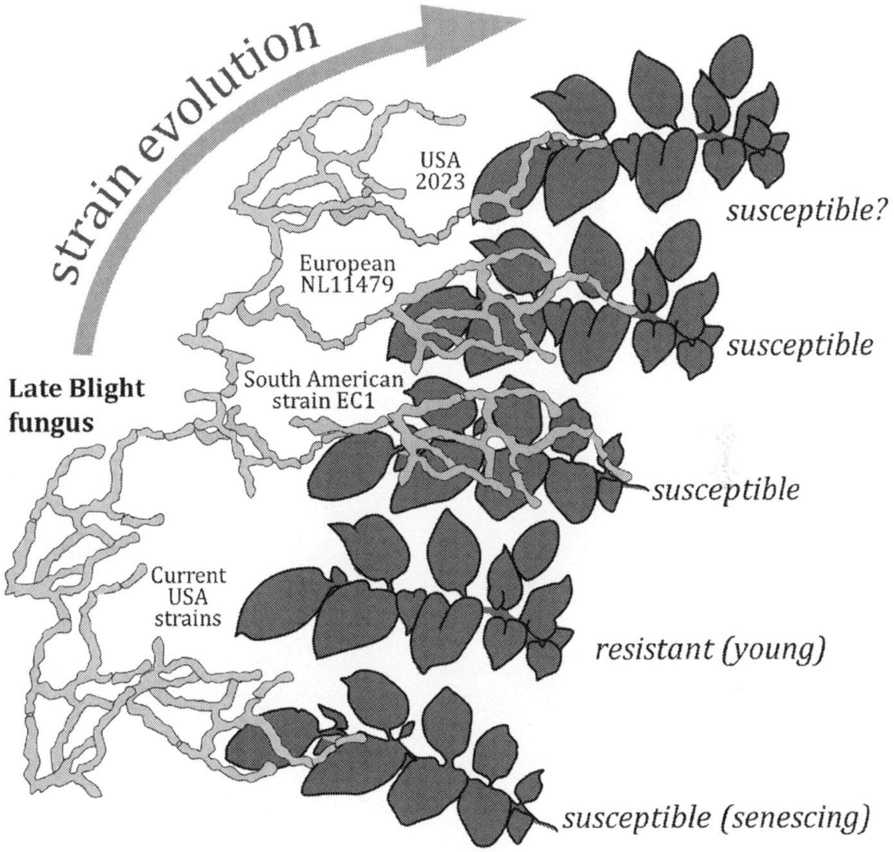
strain evolution
USA
2023
European
NL11479
South American
strain EC1
Late Blight
fungus
Current
USA
strains
susceptible?
susceptible
susceptible
resistant (young)
susceptible (senescing)

5

POTATO ENGINEERING

Compromised Yield

YIELD IS THE MOST IMPORTANT agricultural trait. It is so important that breeders will usually release a new variety only if it offers a yield advantage over existing varieties. Genetic engineers are not so sensible, it appears, because the yield potential of Pandora's Potatoes is severely compromised and much lower than the yield potential of original varieties such as Russet Burbank.

Versions F11, F37, E12, and E24 of Pandora's Potatoes are most accurately evaluated for yield (based on tuber seed that was somewhat adequate, i.e., in part from 2nd generation field-grown plants) and show an average yield drag of **-18.3 percent** [84]. This unacceptable agronomic detriment, hidden behind inappropiate statistics*, would cost farmers about $700 per acre [85].

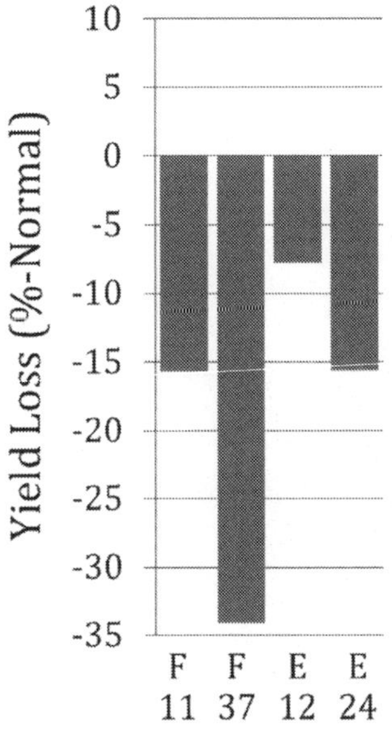

* *As can be confirmed by independent statisticians.*

Preliminary yield determinations of the remaining versions (based on immature seed for inaccurate yields) suggest an average yield drag of **-6.4 percent** [31,58,84] and an associated monetary loss of $250 per acre. I expect higher yield losses for commercial seed. The resulting losses in sale-volumes mostly outweigh the *promised* value of the biotech traits, which is $50 per acre for partial bruise resistance, $50 for the loss of fry-induced darkening, and $50-200 for partial late blight control (dependent on location [60,62].

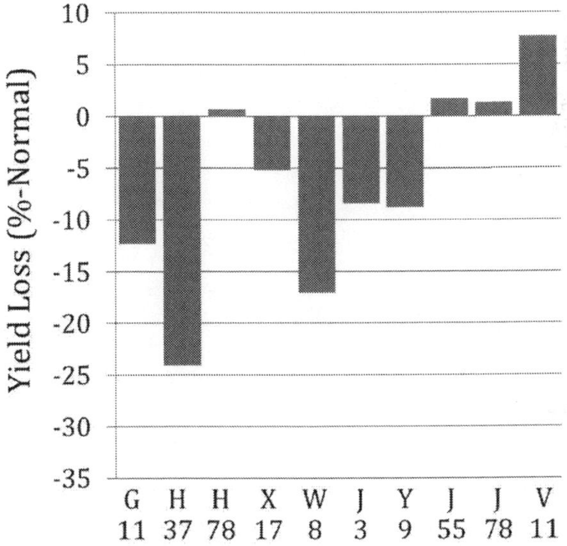

The usable yields are even lower when considering the fact that a significant part of Pandora's Potatoes are compromised by concealed bruise and disease (see Chapter 1).

5

Compromised Size

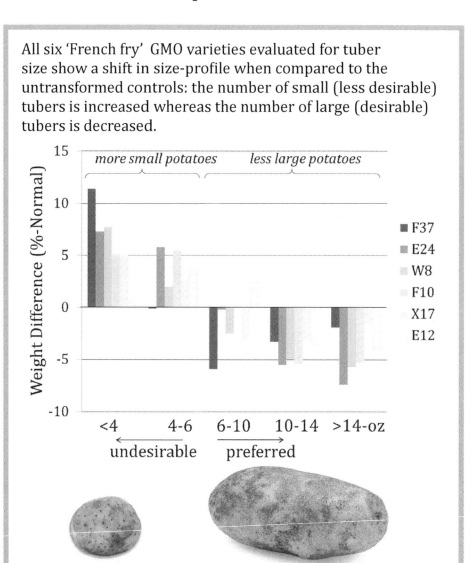

All six 'French fry' GMO varieties evaluated for tuber size show a shift in size-profile when compared to the untransformed controls: the number of small (less desirable) tubers is increased whereas the number of large (desirable) tubers is decreased.

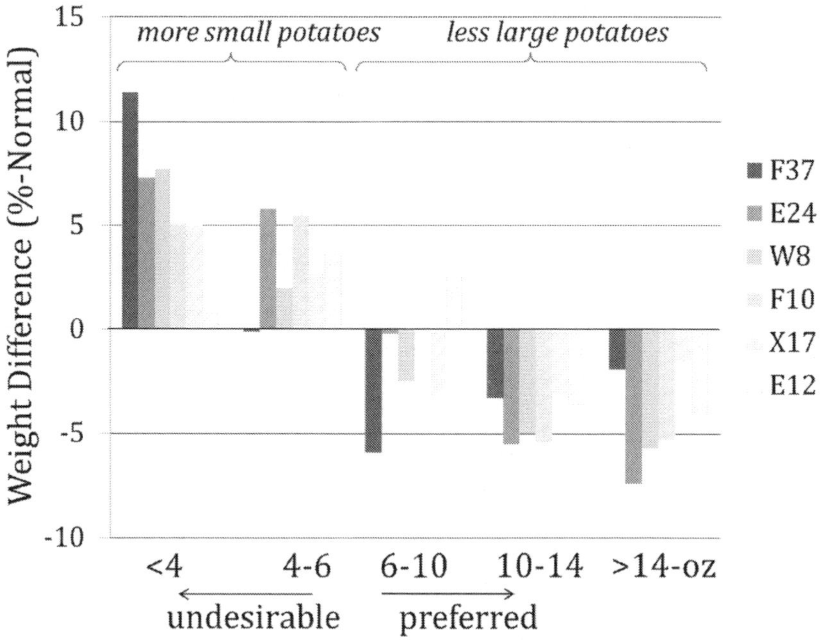

Farmers trying to sell their Pandora's Potatoes may be penalized for the relatively small size, and some loads could even be rejected. For instance, about half the tubers of the varieties E12, E24, and X17 weigh less than 6-oz, which means that processors could impose a five-percent penalty on price, which is equivalent to $200 per acre [**86**].

The Negative Effects on Yield and Size Explained

From 1988 until 2012, I produced about half-a-million different GMO potato 'varieties', and all these 'varieties' were propagated and analyzed for agronomic, molecular, and/or biochemical properties. Some of the modifications were interesting from scientific points of view but the effect on vigor and yield was always negative. My creations were stunted, chlorotic, necrotic, sterile, or altered in shape. And if the issues were not immediately obvious in the greenhouse, they became apparent in the field—everywhere and always.

I learned to manage this **somaclonal variation** [87] by transforming many more plants than I needed. Each DNA construct was introduced into dozens of plants, and only the few lines that seemed to have been affected the least were then propagated and tested. Somaclonal variation affected all modified potatoes, including the NewLeaf varieties developed by Monsanto. According to ex-employees, these varieties suffered a yield drag of *at least* five percent.

Somaclonal variation occurs when the **wrong type of cells** are used to regenerate new plants. These cells are not the superior stem cells meant to produce new plants (such as egg cells and pollen sperm cells) but the inferior somatic cells not intended to outlive the season. Somatic cells from leaves and stems lack the DNA repair systems to prevent the accumulation of mutations, and so they already accumulate mutations even before they are transformed. The number of mutations then dramatically increases when the cells, after transformation, are forced to undergo cancer-like growth.

Each GMO potato contains hundreds to tens-of-thousands of mutations that are not present in the original plant. The mutations occurred at random and their effects are unpredictable. Since about one-third of all mutations affect genes, somaclonal variation is always associated with physiological, biochemical, and phenotypic changes that affect yield [88-94].

GMO potatoes are different from other GMO crops in that, for the following two reasons, transformation-induced mutations cannot be removed through breeding: (1) whenever a GMO variety is crossed with another variety, the thousands of carefully-combined traits of the parents will segregate in the progenies, thus limiting agronomic performance [less than 0.01% of progeny plants will have new combinations of traits that may justify commercialization as new variety, but the identification and development of such a new variety would take at least 15 years of agronomic evaluations and bulk-up and is unlikely to succeed], and (2) whenever a GMO variety is backcrossed with the original variety, the progeny plants will suffer from inbreeding depression.

My conclusion is that potatoes are not amenable to genetic engineering.

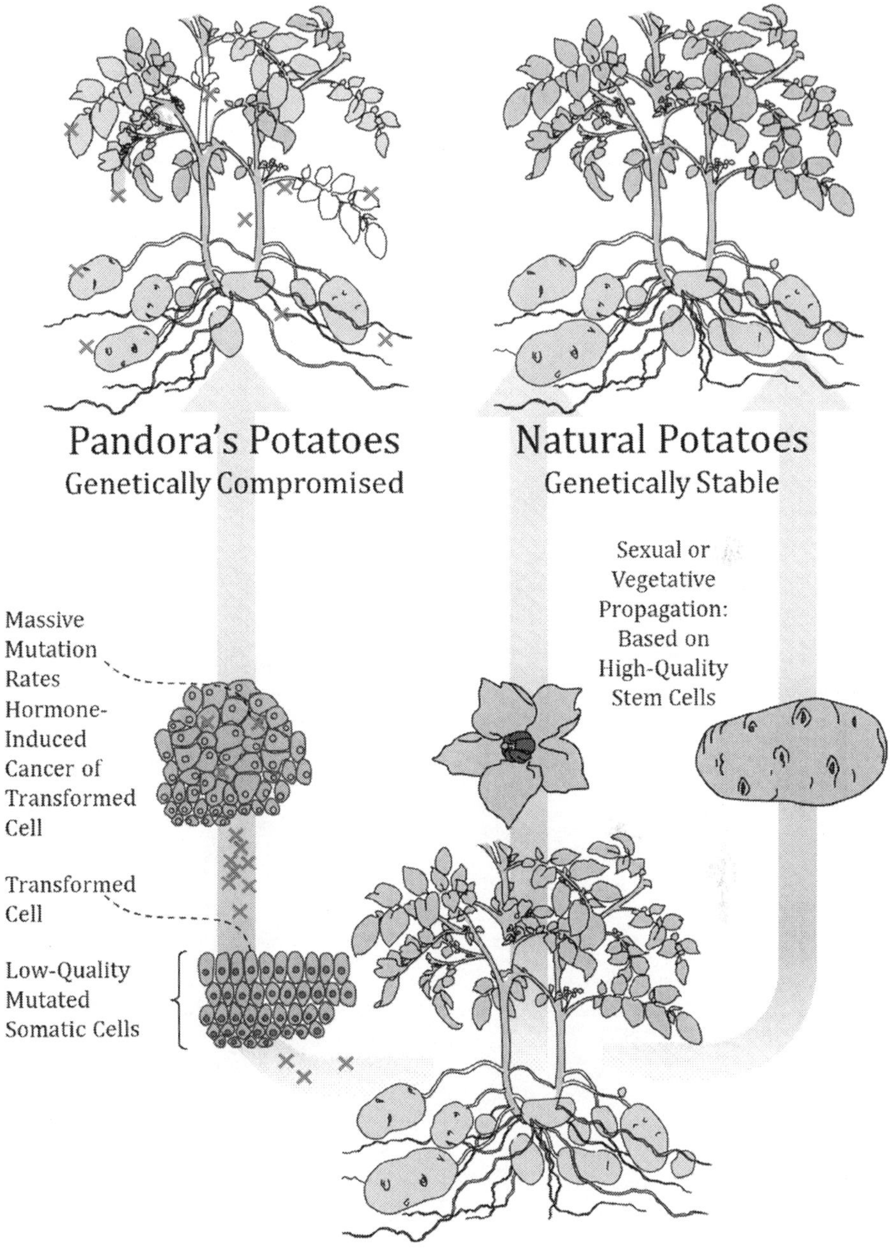
Pandora's Potatoes
Genetically Compromised
Natural Potatoes
Genetically Stable
Sexual or
Vegetative
Propagation:
Based on
High-Quality
Stem Cells
Massive
Mutation
Rates
Hormone-
Induced
Cancer of
Transformed
Cell
Transformed
Cell
Low-Quality
Mutated
Somatic Cells

Trait Instability

Inserted transgenes are often not as stable in plants as the sequences that evolved naturally, especially not when maintained in the heterozygous condition (as is the case in potatoes). Consequently, it is often difficult for breeders at biotech companies to maintain the incorporated traits. There is discomfort about this transgenic instability, and so it represents one of the most under-reported issues regarding GMO crops.

The DNA constructs of Pandora's Potatoes are likely to represent the *most unstable constructs and traits ever commercialized* because they contain numerous hotspots for recombination and epigenetic methylation that genetic engineers usually avoid.

First, the DNA knots used for gene silencing contain very large inverted-repeats. Such structures almost never occur naturally, and they can produce junctions representing key intermediates in homologous recombination when they do [95].

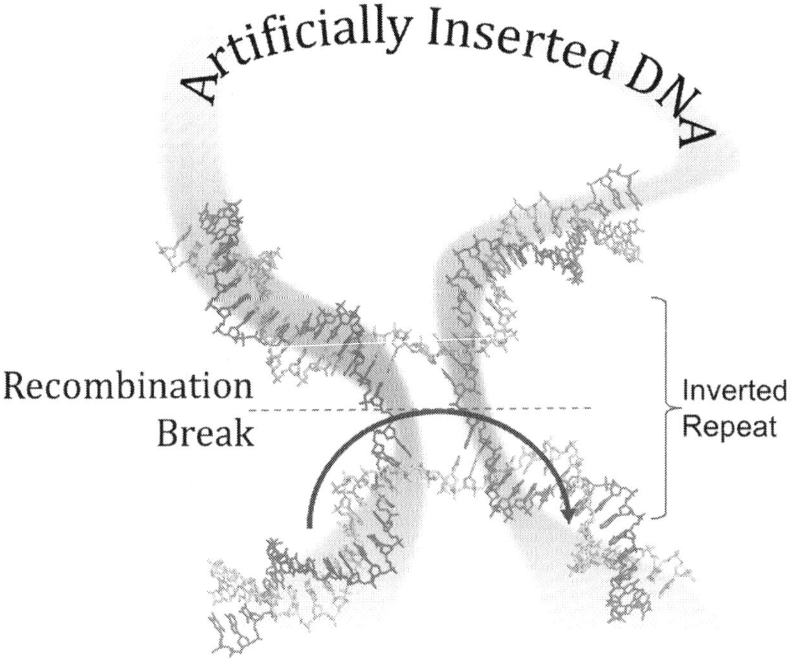

It is therefore to be expected that transgenic inverted repeats cause reversions (through deletions or inversions).

Indeed, Pandora's Potatoes represent the only GMO crop associated with genetic instability *before* commercialization. A small study on the stability of PPO-silencing that was published almost ten years ago showed at least one case of reversion [**84**]:

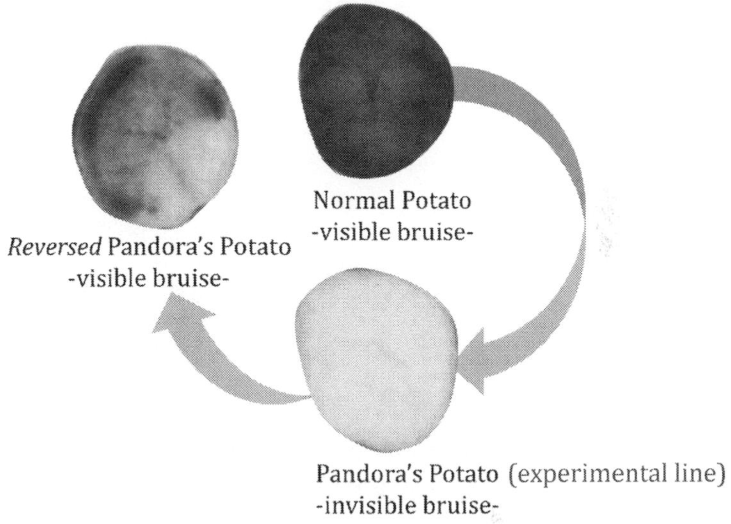

Tuber slices of an untransformed potato (top-right) and two PPO-silenced potatoes (bottom-right and top-left) were treated with catechol to visualize PPO activity. Precipitation of a brown pigment implies PPO activity.

This red flag should have warranted further and more comprehensive examinations, but hardly any data have been disclosed since the early study, and nothing is known about the current frequencies of instability or about the character of new gene fusions resulting from such instability. It is quite discomforting to know that Pandora's Potatoes are propagated and bulked-up in massive amounts but **without an adequate assessment of their instability**.

The **second** source of instability is a group of six transgenic sequences that are identical to twelve native promoter elements**. The elements evolved to drive the expression of genes involved in starch formation, but the function of the transgenic copies was *redirected* to gene silencing.

The resulting increase in the copy number of promoter elements, from twelve in normal potatoes to eighteen in the Pandora version, is like overloading an electric circuit with too many devices, demanding more power than the circuit is able to provide.

And just like the breaker of the overloaded circuit will eventually trip, so is the over-exploitation of genetic promoter elements known to cause *cosuppresion*, i.e., the shutdown of gene expression through DNA methylation [96-99]. Thus, the efficiency of both starch formation and engineered silencing are expected to gradually decline in Pandora's Potatoes.

Two genes that had initially been silenced in Pandora's Potatoes have already become completely ineffective: a construct intended to silence the starch-associated PHL and R1 genes (to prevent the conversion of starch into glucose and fructose) lost its efficacy, and this first example of a 'lost' transgenic trait in a commercial GMO crop is not mentioned anymore in public communications.

The ineffective PHL/R1-silencing construct was substituted (without being removed) by the INV-silencing construct, which also prevents the conversion of starch into glucose and fructose but in a different way.

** *Three engineered copies of the AGP promoter and three engineered copies of the GBSS promoter were introduced into potatoes already containing about eight native copies of the AGP promoter and four native copies of the GBSS promoter.*

An evaluation of published data suggests that the decline in silencing is not limited to PHL and R1: the ASN-silencing construct, which was still relatively effective in the original transformants F10 and J3 appear to have lost about half its activity in the retransformed lines X17 and Y9 [**58**,**84**].

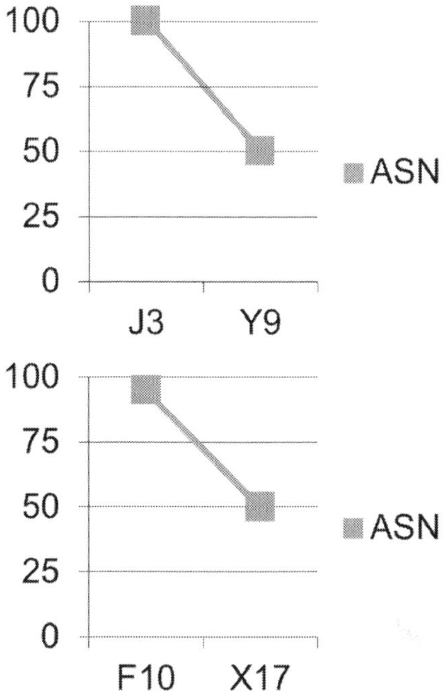

It is to be expected that additional re-transformations will be needed at some time to replace the ASN/PPO-silencing construct as well as the VNT gene, just like how old satellites in space need to be frequently replaced by new ones. The old bits of defunct satellites will forever orbit Earth and so will potatoes forever retain the remnants of junk DNA inserted by genetic engineers.

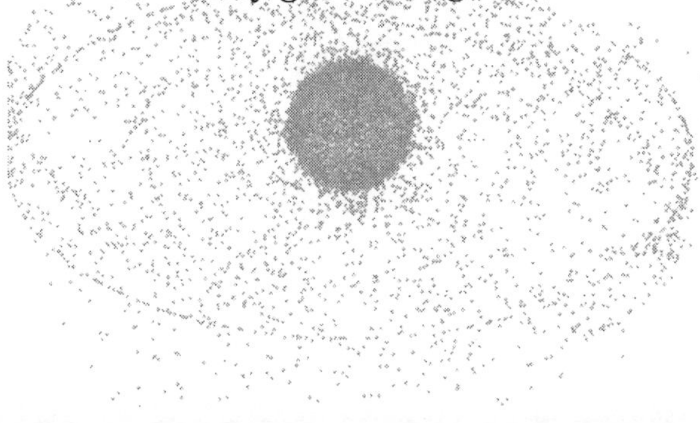

5

Unintended Silencing

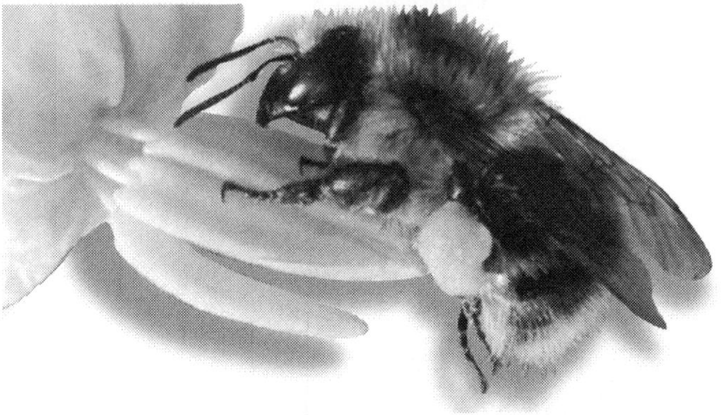

No microarray studies have been carried out to exclude the possibility of any inadvertent silencing of potato's own genes and, more importantly, of the genes of organisms interacting with potato, such as bees, which visit potatoes to collect pollen.

Indeed, the pollen may contain double-stranded RNAs produced by the silencing constructs, even though the silencing constructs seem to have been intended to be 'tuber-specific'. There are no published reports on the extent of PPO/ASN/INV-silencing in pollen, but the petitions for deregulation show that the silencing was effective, unintentionally, in flowers. And the promoters used for silencing are known to be active in pollen.

Thus, it is possible that double-stranded RNAs contaminate the pollen that is collected by nurse bees to produce royal jelly, which is consumed by larvae during the most vulnerable stage of their development. Any ingested cluster bomb fragments will be cleaved into small 20-25 nucleotide fragments that guide the degradation of larval gene transcripts with sequence complementarity, as has been shown previously for other orally-administered double-stranded RNAs [**100-102**].

By comparing the sequence of the double-stranded RNAs generated in potato pollen and the genes that are expressed in bees, some possible candidates for such potential cross-silencing affecting the growth and development of bees are shown on the next page.

<table>
<tr><th>Homology dsRNA - bee RNA</th><th>Predicted function in bees</th></tr>
<tr><td>GGTGCTGACGAAATTTTTGGTGGCT
||||||| ||||||||||||| |||
GGTGCTGCCGAAATTTTTGGTCGCT</td><td>Intraflagellar transport protein 122 homolog</td></tr>
<tr><td>GATTTCATTGAAGTTGTTATTCAAGAATAA
|||||||||||| ||||| ||| ||||||
GATTTCATTGAATTTGTTGTTCGCGAATAA</td><td>Uncharacterized LOC107998915</td></tr>
<tr><td>ATGTTGAAATCAGCTTTGTTGCTTGATTTCATTG
|||||||||||||||| || | |||| ||||
ATGTTGAAATCAGCTTAATTATTCAATTTAATTG</td><td>Uncharacterized LOC110120158</td></tr>
<tr><td>GATGCTATTGAAGATGTTATATATCA
||| | |||||||||||||||| |||
GATCCAATTGAAGATGTTATATTTCA</td><td>Cytochrome oxidase subunit I</td></tr>
<tr><td>TACCATTTCTGGATAAAGAGTTC
||| ||||||||||||||||||
TACGCTTTCTGGATAAAGAGTTC</td><td>Regulator of G-protein signaling 7</td></tr>
<tr><td>TGAAGATGTTATATATCATAT
|||| ||||||||||||||||
TGAATATGTTATATATCATAT</td><td>Uncharacterized variant X3</td></tr>
<tr><td>TATCAGGGGAAGGTGCTGACGAAATTTTT
||||| | ||||||||||||||||||
TATCAAGTTTGGGTGCTGACGAAATTTTT</td><td>Probable mitochondrial trans-2-enoyl-CoA reductase</td></tr>
<tr><td>TTTGTTGCTTGATTTCATTGAAGTT
|| ||| |||||||||||||||||
TTGGTTTATTGATTTCATTGAAGTT</td><td>Uncharacterized LOC107188474</td></tr>
<tr><td>TACTTCCACAAGGCTCCAAAC
|| ||||||||||||||||||
TATTTCCACAAGGCTCCAAAC</td><td>Programmed cell death protein 6 variant X2</td></tr>
<tr><td>ATGCTATTGAAGATGTTATATA
||||||||||||||||| | ||
ATGCTATTGAAGATGTTTTCTA</td><td>Cofactor biosynthesis protein 1 variant X1</td></tr>
<tr><td>ATGTTGAAATCAGCTTTGTTGCTTGATTTCATTG
|||||||||||||||| || | |||| ||||
ATGTTGAAATCAGCTTAATTATTCAATTTAATTG</td><td>Uncharacterized LOC110120158</td></tr>
<tr><td>ATTTCTGGATAAAGAGTTC
|||||||||||||||||||
ATTTCTGGATAAAGAGTTC</td><td>Regulator of G-protein signaling 7</td></tr>
<tr><td>ATGCTATTGAAGATGTTATATA
||||||||||||||||| | ||
ATGCTATTGAAGATGTTTTCTA</td><td>Molybdenum cofactor biosynthesis protein 1 variant X2</td></tr>
</table>

<table>
<tr><th>Homology dsRNA - bee RNA</th><th>Predicted function in bees</th></tr>
<tr><td>CTCTCCTCTTTCCTTTTGCTTTCTGT
||||||||||||||||| | ||| ||
CTCTCCTCTTTCCTTTTCCGTTCCGT</td><td>EST1A-like telomerase-binding protein</td></tr>
<tr><td>CTCCCGTTCGCCGGCGCCGCC
||||||||||||||||| |||
CTCCCGTTCGCCGGCGCTGCC</td><td>Espinas-like protein</td></tr>
<tr><td>GATTCAGCTATTTGGGGAAATATCA
|||||| ||||||||||||||||
GATTCAAGAATTTGGGGAAATATCA</td><td>89 kDa-like centrosomal protein</td></tr>
<tr><td>ATGGTATCATCTTTTTTATC
||| ||||||||||||||||
ATGTTATCATCTTTTTTATC</td><td>UBX domain-containing protein 4-like</td></tr>
<tr><td>ATTGGATGAACGATCCTAATG
|| | ||||||||||||||||
ATCGAATGAACGATCCTAATG</td><td>Uncharacterized variant X2</td></tr>
<tr><td>GTATCATCTTTTTTATCAATACAAT
||| | | ||||||||||||||||
GTAGCTTAATTTTTATCAATACAAT</td><td>Alpha-trehalose-phosphate synthase variant X3</td></tr>
<tr><td>ATACAATCCAGATTCAGCTAT
|||||||||||||||| | ||
ATACAATCCAGATTCAACGAT</td><td>HEAT repeat-containing protein 1</td></tr>
</table>

Monetary Value

A misinformed economist assumed that the use of Pandora's Potatoes would make it possible to omit all late blight spray applications, increase yield (5% less rot), and enhance the efficiency of storage and processing—at a **value of $1,000 per acre** [62].

However, as explained in the previous chapters:

- Some late blight fungicide sprays are still necessary, and the lost tolerance to diseases and pests needs to be compensated for by the increased application of all other pesticides at a cost of $20 per acre in the Northwest to $120 per acre in the Northeast.
- Yields are decreased by at least 6.5% (rather than increased by 5%) at a cost of $200 per acre [85]), and the increase in bruise-associated disease and dehydration during storage may also increase post-harvest losses by $180 per acre [47,61,63].
- Furthermore, the losses due to quality issues and consumer rejection could be insurmountable.

Thus, instead of generating profits, Pandora's Potatoes may cause an average **loss of at least $300 per acre**.

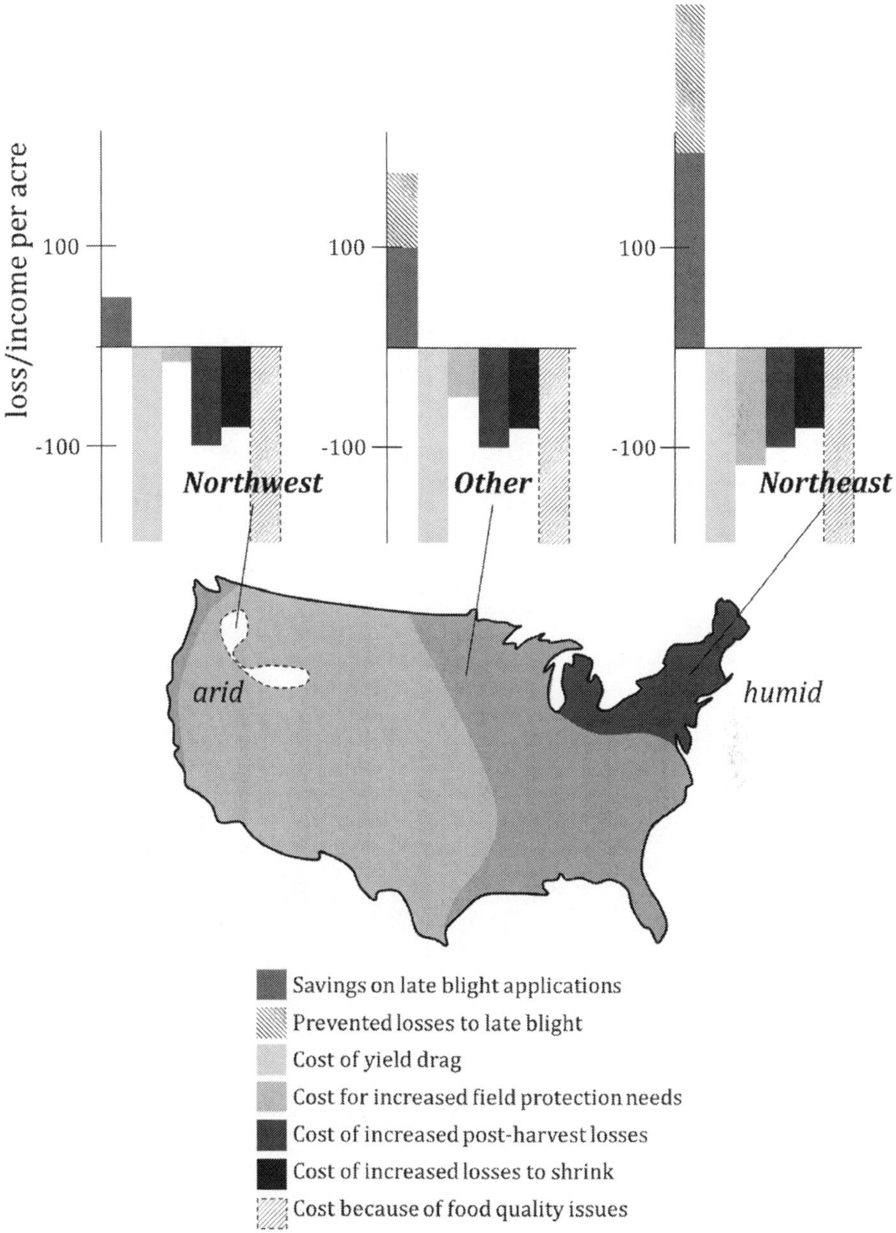

5

Conclusions

Pandora's Potatoes are promoted as superior varieties. Supposedly, they maintained all the traits of the original varieties (agronomics, yield, typeness, and sensory qualities) and gained four new and stable traits: bruise resistance, reduced fry darkening, reduced carcinogenicity, and late blight resistance.

For the following reasons, this description is inaccurate:

1. The two most important traits of normal potatoes were lost: Pandora's Potatoes are compromised in yield potential and produce smaller (less desirable) tubers. Additionally, they emerge later, senesce earlier, and are likely to suffer from other yet unnoticed and/or unreported issues. These undesirable characteristics are mostly evident from data in the Petitions for Deregulation but hidden behind deceptive statistics.
2. The genetic modification did not result in bruise resistance but in a concealment of bruises and infections. This concealment was achieved by eliminating potato's ability to produce protective melanin, which means by increasing potato's vulnerability to infections, pests, environmental stresses, and water loss. The inability to dispose of tubers with concealed bruises and infections will complicate efforts to contain the spread of plant diseases, and it will also result in the exposure of consumers to the toxins that are formed in bruised and infected tissues, such as tyramine, aminoadipate, fungal toxins, and allergens.
3. Instead of admitting the health issues, Pandora's Potatoes are claimed to be less carcinogenic than normal potatoes. However, no supporting data are provided, and normal fries and chips are not carcinogenic to begin with. It appears the health claim is meant to distract consumers from the actual issues, which are (in addition to those associated with bruise and infection) the excess of salt and fat.

4. The claim that Pandora's Potatoes display disease resistance is misleading because the incorporated late blight resistance is unstable and linked to the enhanced vulnerability caused by PPO-silencing. Furthermore, the late blight resistance may entice farmers to grow Pandora's Potatoes where late blight is most aggressive, which is also where numerous other diseases are most aggressive. Thus, any attempt to expand the production of potatoes in these regions will not decrease but increase disease issues. The theoretical possibility that a few versions of Pandora's Potatoes will replace hundreds of natural varieties is worrisome because the resulting increased uniformity will further exacerbate disease and insect issues. Altogether, any temporary savings in late blight fungicide sprays will be dwarfed by the increased expenses for all other input chemicals.
5. The country with sovereign rights over VNT (Argentina) did not authorize the commercial exploitation of its gene by a foreign company.
6. Even though the biotech traits are claimed to be genetically stable, two of them have already become nonfunctional (PHL-silencing and R1-silencing), a third one has been shown to be unstable (PPO-silencing), and at least two more traits are gradually losing efficacy (ASN-silencing and VNT-based resistance to late blight). The presence of large inverted repeats and repeated promoter elements suggests a continued decline in trait efficacy through recombination and methylation of the associated DNA.

Promoted Benefits of Pandora's Potatoes:

Processor Efficiency, Late Blight Resistance, Enhanced Health

Hidden Detriments:

Concealed Bruise & Disease, Concealed Toxins,
Lost Color, Aroma, and Flavor, Yield Drag, Stress Sensitivity,
Reduced Size, and so on.

Because of all the issues,
I see no option but to renounce my creations
and to implore their withdrawal from the market.

In case Pandora's Potatoes are not withdrawn, the following questions should be answered as soon as possible:

- How will farmers be **reimbursed** for losses in yield and tuber quality and for diseases and epidemics originating from infected but symptomless tubers?
- When will claims regarding 'bruise-free' and 'reduced carcinogenicity' be **withdrawn**?
- When will third-party data be **disclosed** regarding the taste profiles and toxin levels?
- How will the Argentinian government be **compensated** for the theft of VNT?

6

A BETTER WAY FORWARDS

PANDORA'S POTATOES were designed to support processors in their continuing effort to **mass-produce uniformity**. They represent yet another step towards the conversion of a once-powerful crop into clonally-propagated 'clumps of starch' that require constant care to survive. Farmers and consumers are expected to pay the price—farmers by providing the daily care for a most vulnerable crop and consumers by eating a food that is robbed of its original wholesomeness.

But there is no excuse to limit potato's diversity and grow billions of identical GMO clones on hundreds of thousands of acres in the United States alone. Potatoes are only good and useful if they are tolerant as a crop and nutritious as a food. And this means that they need to be genetically-diverse, as they still are in parts of South America.

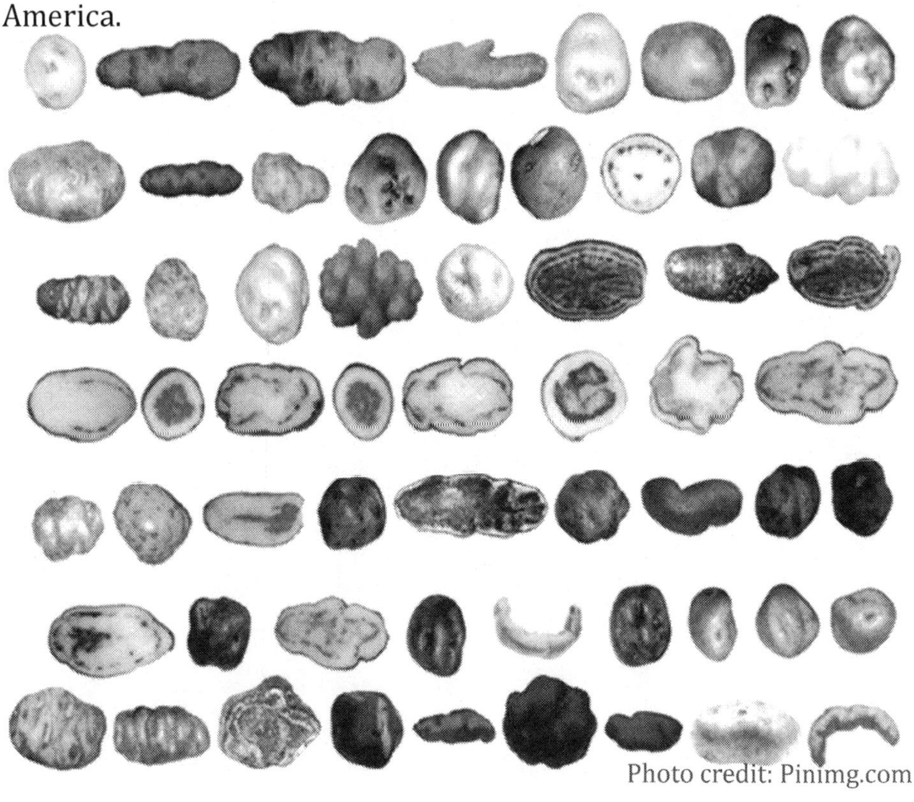

Photo credit: Pinimg.com

Indeed, the Andean farmers know that mixed varieties are more productive and sustainable than single clones—it is an insight confirmed by modern research [103-106]—and so they grow dozens to hundreds of different varieties on single plots, as they have been doing for thousands of years.

Furthermore consumers benefit from both the diversity of tastes and the synergistic health-promoting activities of phytochemicals (such as antioxidant carotenoids and anthocyanins, tocopherol, phytate, folate, and anti-glycemic compounds), most of which are not produced in the GMO potatoes [107].

Potatoes need diversity to thrive, just like people do.

So, instead of increasing the uniformity of potatoes through genetic modification, it is important to start breeding for genetic diversity with methods such as hybrid-seed technology [108,109].

REFERENCES

1. Hensel DR, Locascio SJ (1987) Effect of rates, form, and application date of nitrogen on growth of potatoes. Proc Fla State Hort Soc 100: 203-205.
2. Ohashi M, Ishiyama K, Kojima S, Konishi N, Nakano K, Kanno K, Hayakawa T, Yamaya T (2015) Asparagine synthetase1, but not asparagine synthetase2, is responsible for the biosynthesis of asparagine following the supply of ammonium to rice roots. Plant Cell Physiol 56: 769-778.
3. Wang L, Ruan YL (2016) Critical Roles of Vacuolar Invertase in Floral Organ Development and Male and Female Fertilities Are Revealed through Characterization of GhVIN1-RNAi Cotton Plants. Plant Physiol 171: 405-423.
4. Yan H, Chretien R, Ye J, Rommens CM (2006) New construct approaches for efficient gene silencing in plants. Plant Physiol 141: 1508-18.
5. Peet RC, Mohan-Ram V (2008) Brief on Appeal. December 4, 2008. Retrievable at https://globaldossier.uspto.gov/#/details/US/11233483/A/102596
6. Shepherd LVT (2015) Impacts on the metabolome of down-regulating polyphenol oxidase in potato tubers. Transgenic Research 24: 447-461.36.
7. Yochkova SD, Ivanov EA, Davidova RI (2011) Effect of α-aminoadic acid on Müller cells in retina. J Biomed Clin Res 4: 69-76.
8. Goldberg T, Cai W, Peppa M, Dardaine V, Baliga BS, Uribarri J, Vlassara H (2004) Advanced glycoxidation end products in commonly consumed foods. J Am Diet Assoc 104: 1287-1291.
9. Peppa M, Brem H, Ehrlich P, Zhang JG, Cai W, Li Z, Croitoru A, Thung S, Vlassara H (2003) Adverse Effects of Dietary Glycotoxins on Wound Healing in Genetically Diabetic Mice. Diabetes 52: 2805-2813.
10. Peppa M, He C, Hattori M, McEvoy R, Zheng F, Vlassara H (2003) Fetal or Neonatal Low-Glycotoxin Environment Prevents Autoimmune Diabetes in NOD Mice. Diabetes 52: 1441-1448.
11. Vlassara H, Cai W, Crandall J, Goldberg T, Oberstein R, Dardaine V, Peppa M, Rayfield EJ (2002) Inflammatory mediators are induced by dietary glycotoxins, a major risk factor for diabetic angiopathy. Proc Natl Acad Sci USA 99: 15596-15601.
12. Heijst JWJ, Niessen HWM, Hoekman K, Schalkwijk CG (2005) Advanced Glycation End Products in Human Cancer Tissues: Detection of N-(Carboxymethyl)lysine and Argpyrimidine. Ann NY Acad Sci 1043: 725-733.
13. Llorente B, Alonso GD, Bravo-Almonacid F, Rodríguez V, López MG, Carrari F, Torres HN, Flawiá MM (2010) Safety assessment of nonbrowning potatoes: opening the discussion about the relevance of substantial equivalence on next generation biotech crops. Plant Biotechnol J 9: 136-150.
14. http://foodb.ca/foods/175.
15. Rozan P, Kuo YH, Lambein F (2001) Nonprotein amino acids in edible lentil and garden pea. Amino Acids 20: 319-324
16. http://www.nationalpotatocouncil.org/files/6915/0030/5904/per_capita_utiliz.pdf
17. Cretenet M, Goven J, Heinemann JA, Moore B, Rodriguez-Beltran C (2006) Submission on the DAR for Application A549 food derived from high lysine corn LY039: to permit the use in food of high lysine corn. Submitted to Food Standards Australia/New Zealand (FSANZ), Centre for Integrated Research in Biosafety.
18. "Europe balks at GE corn in NZ", Stuff.co, National, 2 November 2009, http://www.stuff.co.nz/national/3020246/Europe-balks-at-GE-corn-in-NZ.
19. Swiedrych A, Stachowiak J, Szopa J (2004) The catecholamine potentiates starch mobilization in transgenic potato tubers. Plant Physiol Biochem 42: 103-109.
20. Borg-Olivier O, Monties B (1993) Lignin, suberin, phenolic acids and tyramine in the suberized, wound-induced potato periderm. Phytochemistry 32: 601–606.
21. Guillet G, De Luca V (2005) Wound-inducible biosynthesis of phytoalexin hydroxycinnamic acid amides of tyramine in tryptophan and tyrosine decarboxylase transgenic tobacco lines. Plant Physiol 137: 692-9.
22. Dao L, Friedman M (1992) Chlorogenic acid content of fresh and processed potatoes deter-

mined by ultraviolet spectrophotometry. J Agric Food Chem 40: 2152-2156.
23. Manual of Clinical Dietetics. 6th ed. Chicago, Il. American Dietetic Association; 2000.
24. Garcia E, Santos C (2017) Toxicity, Monoamine Oxidase Inhibitor. StatPearls. Treasure Island (FL): StatPearls Publishing.
25. Sullivan ML (2015) Beyond brown: polyphenol oxidases as enzymes of plant specialized metabolism. Front Plant Sci 5: 783.
26. Hunt LT, George DG, Yeh LS (1985) Ragweed allergen Ra3: relationship to some type 1 copper-binding proteins. J Mol Evol 21: 126-32.
27. Webb KJ, Cookson A, Allison G, Sullivan ML, Winters AL (2014) Polyphenol oxidase affects normal nodule development in red clover (Trifolium pratense L.). Front Plant Sci 17: 700.
28. Araji S, Grammer TA, Gertzen R, Anderson SD, Mikulic-Petkovsek M, Veberic R, Phu ML, Solar A, Leslie CA, Dandekar AM, Escobar MA (2014) Novel roles for the polyphenol oxidase enzyme in secondary metabolism and the regulation of cell death in walnut. Plant Physiol 164: 1191-1203.
29. Rommens CM, Ye J, Richael C, Swords K (2006) Improving potato storage and processing characteristics through all-native DNA transformation. J Agric Food Chem 54:9882-9887.
30. Hakimi S, Krohn BM, Stark DM (2006) Method of imparting disease resistance in plants by reducing PPO levels. US Patent 7,122,719 B2.
31. Clark P, Habig J, Ye J, Collinge S (2014). Petition for determination of nonregulated status for Innate potatoes. No. 14-093-01p-Aphis-USDA.
32. Boeckx T, Winters A, Webb KJ, Kingston-Smith AH (2017) Detection of Potential Chloroplastic Substrates for Polyphenol Oxidase Suggests a Role in Undamaged Leaves. Front Plant Sci 8: 237.
33. Ngazee E, Icishahayo D, Coutinho TA, Van der Waals JE (2012) Role of polyphenol oxidase, peroxidase, phenylalanine ammonia lyase, chlorogenic acid, and total soluble phenols in resistance of potatoes to soft rot. Plant Dis 96: 186–192.
34. Castañera P, Steffens JC, Tingey WM (1996) Biological performance of Colorado potato beetle larvae on potato genotypes with differing levels of polyphenol oxidase. J Chem Ecol 22: 91–101.
35. Vanitha SC, Niranjana SR, Umesha S (2009) Role of phenylalanine ammonia lyase and polyphenol oxidase in host resistance to bacterial wilt of tomato. J Phytopathol 157: 552–557.
36. Kavitha R., Umesha S (2008) Regulation of defense-related enzymes associated with bacterial spot resistance in tomato. Phytoparasitica 36: 144–159.
37. Li L, Steffens JC (2002) Overexpression of polyphenol oxidase in transgenic tomato plants results in enhanced bacterial disease resistance. Planta 2: 239–247.
38. Thipyapong P, Melkonian J, Wolfe DW, Steffens JC (2004) Suppression of polyphenol oxidases increases stress tolerance in tomato. Plant Sci 167: 693–703.
39. Wang J, Constabel CP (2004) Polyphenol oxidase overexpression in transgenic Populus enhances resistance to herbivory by forest tent caterpillar (Malacosoma disstria) Planta 220: 87–96.
40. Niranjan RS, Sarosh BR, Shetty HS (2006) Induction and accumulation of polyphenol oxidase activities as implicated in development of resistance against pearl millet downy mildew disease. Funct Plant Biol 33: 563–571.
41. Mohammadi M, Kazemi H (2002) Changes in peroxidase and polyphenol oxidase activities in susceptible and resistant wheat heads inoculated with Fusarium graminearum and induced resistance. Plant Sci 162: 491–498
42. Boeckx T, Webster R, Winters AL, Webb KJ, Gay A, Kingston-Smith AH (2015) Polyphenol oxidase-mediated protection against oxidative stress is not associated with enhanced photosynthetic efficiency. Ann Bot 116: 529-40.
43. Araji S, Grammer TA, Gertzen R, Anderson SD, Mikulic-Petkovsek M, Veberic R, Phu ML, Solar A, Leslie CA, Dandekar AM (2014) Novel roles for the polyphenol oxidase enzyme in secondary metabolism and the regulation of cell death in walnut. Plant Physiol 164: 1191–1203.
44. Ahktar W, Mahmoo T (2017) Response of rice polyphenol oxidase promoter to drought and salt stress. Pak J Bot 49: 21-23.
45. Small T (1945) The effect of disinfecting and bruising seed potatoes on the incidence of dry rot (Fusarium caeruleum (Lib) Sacc) Ann Appl Biol 32: 310–318.
46. Olsen N, Miller J, Nolte P (2013) Diagnosis & management of potato storage diseases. University of Idaho. CIS 1131.

47. Thornton M Bohl W (1998) Preventing potato bruise damage. Cooperative Extension System Agricultural Experiment Station. BUL 725 (Revised).
48. EFSA CONTAM (2015) Scientific Opinion on acrylamide in food. EFSA Journal 13: 4104, 321 pp. doi:10.2903/j.efsa.2015.4104.
49. Friedman G (1996) Food browning and its prevention. J Agric Food Chem 44: 631-653
50. Virk-Baker MK, Nagy TR, Barnes S, Groopman J (2014) Dietary Acrylamide and Human Cancer: A Systematic Review of Literature. Nutr Cancer 66: 774–790.
51. Liska DJ, Cook CM, Wang DD, Szpylka J (2016) Maillard reaction products and potatoes: have the benefits been clearly assessed? Food Sci Nutr 4: 234–249.
52. Igarashi D, Ishizaki T, Totsuka K, Ohsumi C (2009) ASN2 is a key enzyme in asparagine biosynthesis under ammonium sufficient conditions. Plant Biotechnology 26: 153-159.
53. Kizito KF, Abdel-Aal MH, Ragab MH, et al (2017) Quality attributes of French fries as affected by different coatings, frozen storage and frying conditions. J Agric Sci Bot 1: 18-24.
54. Oana-Viorela N, Şerban AM, Angheluţă M, Botez E, Andronoiu DG, Mocanu GD (2014) The effects of frying on potatoes physico-chemical, sensory and textural properties. Agricultura 3: 36-42.
55. Sturm A (1999) Invertases. Primary structures, functions, and roles in plant development and sucrose partitioning. Plant Physiol 121: 1-8.
56. Kim J-Y, Mahé A, Brangeon J, Prioul J-L (2000) A Maize Vacuolar Invertase, IVR2, Is Induced by Water Stress. Organ/Tissue Specificity and Diurnal Modulation of Expression. Plant Physiology 124: 71-84.
57. Nägele T, Henkel S, Hörmiller I, Sauter T, Sawodny O, Ederer M, Heyer AG (2010) Mathematical modeling of the central carbohydrate metabolism in Arabidopsis reveals a substantial regulatory influence of vacuolar invertase on whole plant carbon metabolism. Plant Physiol 153: 260-272.
58. Pence M, Spence R, Rood T, Habig J, Collinge S (2016) Petition for extension of nonregulated status for X17 and Y9 potatoes. No. 16-064-01p-Aphis-USDA.
59. Wang L, Ruan YL (2016) Critical Roles of Vacuolar Invertase in Floral Organ Development and Male and Female Fertilities Are Revealed through Characterization of GhVIN1-RNAi Cotton Plants. Plant Physiol 171: 405-423.
60. Halterman D, Guenthner J, Collinge S (2016) Biotech potatoes in the 21st century: 20 years since the first biotech potato. Am. J. Potato Res 93: 1.
61. Guenther JF, Michael KC,Nolte P (2001) The economic impact of potato late blight on U.S. growers. Potato Res 44:121-125.
62. Guenther J (2017) Economic and Environmental Benefits of Biotech Potatoes with Traits for Bruise Resistance, Late Blight Resistance, and Cold Storage. AgBioForum 20: 37-45.
63. Eborn B, 2017, Potato Cost of Production for Idaho. Idaho Potato Commission.
64. Haverkort AJ, Struik PC, Visser RGF, Jacobsen E (2009) Applied biotechnology to combat late blight in potato caused by phytophthora infestans. Potato Research 52: 249–264.
65. Wolfe DW, Comstock J, Menninger H, Weinstein D, Sullivan K, Kraft C, Chabot B, Curtis P, Leichenko R, Vancura P (2011) Responding to Climate Change in New York State: The ClimAID Integrated Assessment for Effective Climate Change Adaptation Final Report. Annals of the New York Academy of Sciences, 1244.
66. www.ofc.org.uk/files/ofc/papers/havenbaker-ppt.pdf.
67. http://pmep.cce.cornell.edu/fqpa/crop-profiles/potato.html.
68. Phillips SL (2002) Variety mixtures and the blighted organic potato. In Proceedings of the BCPC Conference – Pests & Diseases, Eds British Crop Protection Council Brighton, pp. 737–740. Farnham, Surrey: British Crop Protection Council.
69. Wolfe MS (1985) The current status and prospects of multiline cultivars and variety mixtures for disease resistance. Ann Rev Phytopathol 23: 251–273.
70. Akem C, Ceccarelli S, Erskine W, Lenne J (2000) Using genetic diversity for disease resistance in agricultural production. Outlook on Agriculture 29: 25–30.
71. Malcomsom JF (1969) Factors involved in resistance to blight in assessment of resistance using detached leaves. Ann Appl Biol 64: 461-468.
72. Umaerus V, Umaerus M, Erjefält L, Nilsson BA (1983) Control of Phytophthora by host resistance: problems and progress. In: Erwin DS, Barnicki-Garcia S, Tsao PH (eds), Phytophthora: Its

biology, taxonomy, ecology and pathology. APS, St Paul Minnesota: 315-326.
73. Fry W (2015) Late blight update 2015. http://www.hort.cornell.edu/expo/ proceedings/2016/Potato.%20Late%20blight%20update% 202015.%20Fry.pdf
74. Vanderplank JE (1968) Disease Resistance in Plants. Academic Press. New Managing Global Genetic Resources: Agricultural Crop Issues and Policies. Washington, DC: The National Academies Press.
75. Robinson RA (1987) Host management in crop pathosystems. Macmillan, Ne.
76. Song J, Bradeen JM, Naess SK, Raasch JA, Wielgus SM, Haberlach GT, Liu J, Kuang H, Austin-Phillips S, Buell CR, Helgeson JP, Jiang J (2003) Gene RB cloned from Solanum bulbocastanum confers broad spectrum resistance to potato late blight. Proc Natl Acad Sci USA 100: 9128-9133.
77. Helgeson JP, Pohlman JD, Austin S, Haberlach GT, Wielgus SM, Ronis D, Zambolim L, Tooley P, McGrath JM, James RV (1998) Somatic hybrids between Solanum bulbocastanum and potato: a new source of resistance to late blight. Theor Appl Genet 96:738–742.
78. Jones JDG, Witek K, Verweij W, Jupe F, Cooke D, Dorling S, Tomlinson L, Smoker M, Perkins S, Foster S (2014) Elevating crop disease resistance with cloned genes. Philos Trans R Soc Lond B Biol Sci. 369: 20130087.
79. Haverkort AJ, Boonekamp PM, Hutten R, Jacobsen E, Lotzl LAP, Kessel GJT, Vossen JH, Visser RGF (2016) Durable late blight resistance in potato through dynamic varieties obtained by cisgenesis: scientific and societal advances in the DuRPh project. Potato Research 59: 35–66.
80. Foster SJ, Park TH, Pel M, Brigneti G, Sliwka J, Jagger L, van der Vossen E, Jones JDG (2009) Rpi-vnt1.1, a Tm-2 Homolog from Solanum venturii, Confers Resistance to Potato Late Blight. Molecular Plant-Microbe Interactions 22: 589-600.
81. Pel MA, Foster SJ, Park TH, Rietman H, van Arkel G, Jones JD, Van Eck HJ, Jacobsen E, Visser RG, Van der Vossen EA (2009) Mapping and cloning of late blight resistance genes from Solanum venturii using an interspecific candidate gene approach. Mol Plant Microbe Interact 22: 601-615.
82. Zhu S, Vossen JH, Bergervoet MJE, Nijenhuis M, Kodde LP, Kessel GJT, Vleeshouwers VGGA, Visser RGF, Jacobsen E (2014) An updated conventional- and a novel GM potato late blight R gene differential set for virulence monitoring of phytophthora infestans. Euphytica 202: 219–234.
83. Jones JDG, Witek K, Verweij W, Jupe F, Cooke D, Dorling S, Tomlinson L, Smoker M, Perkins S, Foster S (2014) Elevating crop disease resistance with cloned genes. Philos Trans R Soc Lond B Biol Sci. 369: 20130087.
84. Clark P, Collinge S (2013) Petition for determination of nonregulated status for Innate potatoes. No. 13-022-01p-Aphis-USDA.
85. USDA Potatoes 2016 Summary (2017). ISSN: 1949-1514.
86. Bolotova Y, Patterson PE (2009) An Analysis of Contracts in the Idaho Processing Potato Industry. Journal of Food Distribution Research 40: 32-38.
87. Jiang C, Mithani A, Gan X, Belfield EJ, Klingler JP, Zhu JK, Ragoussis J, Mott R, Harberd NP (2011) Regenerant Arabidopsis lineages display a distinct genome-wide spectrum of mutations conferring variant phenotypes. Curr Biol 21: 1385-1390.
88. Zhang D, Wang Z, Wang N, Gao Y, Liu Y, Wu Y, Bai Y, Zhang Z, Lin X, Dong Y, Ou X, Xu C, Liu B (2014) Tissue culture-induced heritable genomic variation in rice, and their phenotypic implications. PLoS One. 9: e96879.
89. Phillips RL, Kaeppler SM, Olhoft P (1994) Genetic instability of plant tissue cultures: breakdown of normal controls. PNAS USA 91: 5222-5226.
90. Brown DCM, Thorpe TA (1995). Crop improvement through tissue culture. World Journal Microbiol Biotechnol 11: 409-415.
91. Hirochika H, Sugimoto K, Otsuki Y, Tsugawa H, Kanda M (1996) Retrotransposons of rice involved in mutations induced by tissue culture. PNAS USA 93: 7783-7788.
92. Kaeppler SM, Kaeppler HF, Rhee Y (2000) Epigenetic aspects of somaclonal variation in plants. Plant Mol Biol 43: 179-88.
93. Jain SM (2001) Tissue culture-derived variation in crop improvement. Euphytica 118: 153-166.
94. Bregitzer P, Zhang S, Cho MJ, Lemaux PG (2002) Reduced somaclonal variation in barley is associated with culturing highly differentiated, meristematic tissue. Crop Science 42: 1303-1308.
95. Wang Y, Leung FC (2006) Long inverted repeats in eukaryotic genomes: recombinogenic motifs determine genomic plasticity. FEBS Lett 580: 1277-84.

96. Stam M, Mol JNM, Kooter JM (1997) The silence of genes in transgenic plants. Ann Bot 79: 3–12.
97. Mette MF, van der Winden J, Matzke MA, Matzke AJ (1999) Production of aberrant promoter transcripts contributes to methylation and silencing of unlinked homologous promoters in trans. EMBO J 18: 241-248.
98. Nocarova E, Opatrny Z, Fischer L (2010) Successive silencing of tandem reporter genes in potato (Solanum tuberosum) over 5 years of vegetative propagation. Ann Bot 106: 565-572.
99. Rajeevkumar S, Anunanthini P, Sathishkumar R (2015) Epigenetic silencing in transgenic plants. Front Plant Sci 6: 693.
100. Desai SD, Eu YJ, Whyard S, Currie RW.. Reduction in deformed wing virus infection in larval and adult honey bees (Apis mellifera L.) by double-stranded RNA ingestion. Insect Mol Biol 2012; 21: 446-55.
101. Hunter W, Ellis J, Vanengelsdorp D, Hayes J, Westervelt D, Glick E, Williams M, Sela I, Maori E, Pettis J, Cox-Foster D, Paldi N (2010) Large-scale field application of RNAi technology reducing Israeli acute paralysis virus disease in honey bees (Apis mellifera, Hymenoptera: Apidae). PLoS Pathog 6: e1001160.
102. Nunes FMF, Simões ZLP (2009) A non-invasive method for silencing gene transcription in honeybees maintained under natural conditions. Insect Biochem Molec 39: 157-160.
103. Hooper DU et al (2005) Effects of biodiversity on ecosystem functioning: a consensus of current knowledge. Ecol Monogr 75: 3-35.
104. Cardinale BJ et al (2012) Biodiversity loss and its impact on humanity. Nature 486: 59-67
105. Gross K et al (2014) Species richness and the temporal stability of biomass production: an analysis of recent biodiversity experiments Am Nat 183: 1-12.
106. Litrico I, Violle C (2015) Diversity in Plant Breeding: A New Conceptual Framework. Trends Plant Sci 20: 604-613.
107. Drescher LS, Thiele S, Mensink GB (2007) A new index to measure healthy food diversity better reflects a healthy diet than traditional measures. J Nutr 137: 647-651.
108. Frison EA (2016) From uniformity to diversity: A paradigm shift from industrial agriculture to diversified agroecological systems. International Panel of Experts on Sustainable Food Systems (IPES-Food).
109. http://solynta.com/

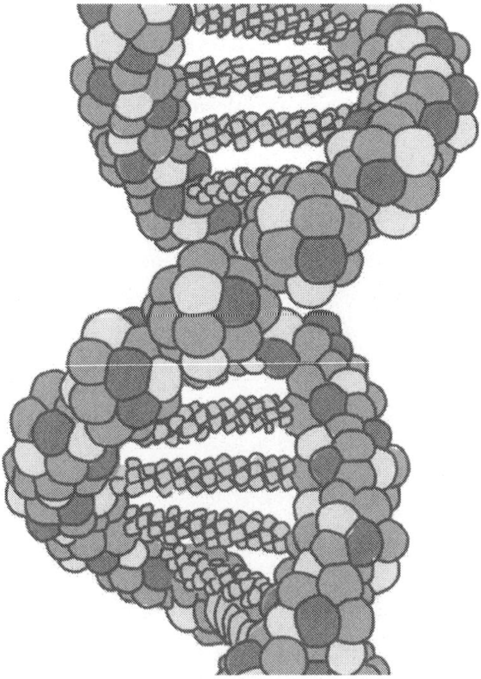

ABOUT THE AUTHOR

I grew up as a 'product' of my time in that I believed that controlled predictability was better than natural capriciousness—society needed to be liberated from the erratic forces of nature!

Ignoring my naive interests in art and religion, I focused on science instead. It was the right thing to do, I was told. By mindlessly enacting the well-meant advice of others, I lost myself and ended up in the laboratory, extracting DNA from dead tissues, modifying this DNA in bacteria, and then transforming plants with the resulting constructs.

For 25 years, I tried to realize the promise of genetic engineering and ignored the detriments of my work. My final achievement was Pandora's Potato, which I created for one of the largest potato processors in the world. The relevant methods and products were published in about one-hundred scientific papers and patents.

It took me too much time to acknowledge that genetic engineering is not a true science or even a profession but the expression of a distorted mindset. Unable to see, touch, hear, smell, or taste DNA, I used guesses, assumptions, theories, and dogmas to develop a shadow image of the dead and fragmented essence of life as something that could be altered in any way imaginable. I could undo whatever evolution had established over the course of millions of years of life-and-death experiences in a single day.

Ultimately, I renounced my past work and refocused my efforts on traditional forms of breeding to develop genetically-diverse crops for the production of tasty and healthy foods. I also try to address the lingering issues caused by my work by clarifying the hidden issues of genetic engineering.

I currently live in France.

96987895R00052

Made in the USA
Middletown, DE
02 November 2018